MATH OLYMPIAD ALGEBRA

MATH OLYMPIAD ALGEBRA

Roman Kvasov, Ph.D.

For general information on our other products and services please contact our Customer Service at info@42points.com.

42 Points also publishes its books in a variety of electronic formats. Some content that appears in print, however, may not be available in electronic format.

MATH OLYMPIAD ALGEBRA / Roman Kvasov, Ph.D.
Printed in the United States of America.

10 9 8 7 6 5 4 3 2 1

Contents

Dedicated to my wife Arlenys

Introduction

Every mathematics competition has its own unique format, set of rules, and most importantly – the list of topics that it covers. Mastering these topics often guarantees outstanding performance on the competition. This book provides an introduction to the most popular topics, ideas, and techniques used in math olympiad algebra. It also includes 150 practice problems with full solutions.

The book is structured into 50 chapters, each dedicated to a specific topic or technique. All chapters begin with a brief discussion of the main idea and then immediately delve into practice problems. Some of the problems in this book are well-known, some are modifications of existing problems, and the majority are original and come from the author's personal archive accumulated over 20 years of math olympiad coaching.

Although the book covers fundamental topics in school Algebra, Precalculus, and Calculus, it is not a substitute for a textbook and does not intend to be one. It is rather a well-organized, extensive review of the types of problems that appear on mathematics competitions year after year. For students who are just beginning their math olympiad journey, the book will serve as a good introductory material. More experienced students will find the book particularly useful a few weeks before the competition, when they want to review topics and work through problems. In either case, the book will serve as a good indicator of which topics should be practiced more.

It is important to note that preparing for any mathematics competition is a long journey that requires a lot of time and effort. In addition to knowing the theory, students should also include a significant amount of practice in their daily preparation. After finishing this book, it is highly recommended to work through all available past exams of the math olympiad that you are preparing for. This will complete your preparation and boost your confidence.

The author wishes you the best of luck on your math olympiad journey and hopes that you will enjoy the book.

Roman Kvasov, Ph.D.

CHAPTER 1

Common Factor

Many math olympiad problems rely on the **common factor**, which is the process of factoring out the largest repeating term from a set of algebraic expressions. This fundamental factorization technique is used to simplify expressions, solve equations, and work with polynomials.

Common factor factorization is based on the **Inverted Distributive Properties**.

Inverted Distributive Properties

$$ab + ac = a(b + c)$$
$$ab - ac = a(b - c)$$

Let us now consider several problems.

Problem 1

Given that
$$2^{n+2} + 2^{n+1} + 2^n = 2^{m+3} - 2^m$$
Prove that $n = m$.

Solution

Let us rewrite the equation and factor out the common factor on both sides:
$$2^{n+2} + 2^{n+1} + 2^n = 2^{m+3} - 2^m$$
$$2^n \cdot 4 + 2^n \cdot 2 + 2^n = 2^m \cdot 8 - 2^m$$
$$2^n \cdot (4 + 2 + 1) = 2^m \cdot (8 - 1)$$
$$2^n \cdot 7 = 2^m \cdot 7$$
$$2^n \cdot \cancel{7} = 2^m \cdot \cancel{7}$$
$$2^n = 2^m$$
$$2^{n-m} = 1$$

which implies that $n - m = 0$, or equivalently $n = m$, as desired.

Problem 2

Positive numbers a, b and c satisfy the equalities
$$a + b = c$$
$$a^3 + a^2b + a^2 = c + 1$$
Prove that at least one of the numbers is equal to 1.

Solution

It will be enough to prove that $a = 1$.

Let us factor out the common factor a^2 in the second equation and substitute $a + b$ for c from the first equation:
$$a^3 + a^2b + a^2 = c + 1$$
$$a^2(a + b + 1) = c + 1$$
$$a^2(c + 1) = c + 1$$
$$a^2(c + 1) - (c + 1) = 0$$

Let us now factor out the common factor $(c + 1)$:

$$a^2(c + 1) - (c + 1) = 0$$
$$(c + 1)\left(a^2 - 1\right) = 0$$
$$(c + 1)(a - 1)(a + 1) = 0$$

Notice that since a and c are positive numbers, then the last equality implies that $a = 1$, as desired.

Problem 3

Real numbers x, y and z satisfy the equations:

$$xyz = 1$$
$$x + y + z = 1$$

Prove the equality

$$(xy + xz - x)(yz + xy - y)(xz + yz - z) = -1$$

Solution

Start by noticing that from the equation $x + y + z = 1$ we have

$$x + y - 1 = -z$$
$$y + z - 1 = -x$$
$$z + x - 1 = -y$$

Let M denote the expression on the left-hand side of the needed equality. Let us factor out the common factor in each parenthesis of the expression M. We have

$$
\begin{aligned}
M &= (xy + xz - x)(yz + xy - y)(xz + yz - z) \\
&= x(y + z - 1)\, y(z + x - 1)\, z(x + y - 1) \\
&= xyz(y + z - 1)(z + x - 1)(x + y - 1) \\
&= (1)(y + z - 1)(z + x - 1)(x + y - 1) \\
&= (-x)(-y)(-z) \\
&= -xyz \\
&= -1
\end{aligned}
$$

as desired.

CHAPTER 2

Difference of Squares Formula

Many math olympiad problems rely on the **Difference of Squares Formula**. This formula is often used to factor expressions and efficiently simplify lengthy products.

Difference of Squares Formula

$$a^2 - b^2 = (a - b)(a + b)$$

Let us consider several problems.

Problem 1

Distinct real numbers a and b are such that

$$\frac{a}{b} + a = \frac{b}{a} + b$$

Find the value of

$$\frac{1}{a} + \frac{1}{b}$$

Solution

Answer: -1.

Let us pass all the terms to the left-hand side and use the Difference of Squares Formula:

$$\frac{a}{b} + a = \frac{b}{a} + b$$

$$\frac{a}{b} - \frac{b}{a} + a - b = 0$$

$$\frac{a^2}{ab} - \frac{b^2}{ab} + a - b = 0$$

$$\frac{a^2 - b^2}{ab} + a - b = 0$$

$$\frac{(a-b)(a+b)}{ab} + a - b = 0$$

Let us now factor the common factor[1] of $a - b$:

$$\frac{(a-b)(a+b)}{ab} + a - b = 0$$

$$(a-b)\left(\frac{a+b}{ab} + 1\right) = 0$$

Since a and b are distinct, then $a - b \neq 0$. Therefore

$$\frac{a+b}{ab} + 1 = 0$$

$$\frac{a+b}{ab} = -1$$

$$\frac{a}{ab} + \frac{b}{ab} = -1$$

$$\frac{1}{b} + \frac{1}{a} = -1$$

$$\frac{1}{a} + \frac{1}{b} = -1$$

as desired.

---

[1] You can find more information about this topic in Chapter 1 "Common Factor".

Problem 2

Show that there are no positive real numbers that satisfy both equations

$$x = y + 1$$
$$x^{16} = (x + y) \cdot (x^2 + y^2) \cdot (x^4 + y^4) \cdot (x^8 + y^8)$$

Solution

Let us assume that such positive real numbers exist.

Notice that $x = y + 1$ implies that $x - y = 1$. Therefore, if we multiply the right-hand side of the second equation by $(x - y)$, it will not change its value. Nevertheless, this artificial multiplication will allow us to apply the Difference of Squares Formula.

We have

$$\begin{aligned}
x^{16} &= (x + y) \cdot (x^2 + y^2) \cdot (x^4 + y^4) \cdot (x^8 + y^8) \\
&= (x - y) \cdot (x + y) \cdot (x^2 + y^2) \cdot (x^4 + y^4) \cdot (x^8 + y^8) \\
&= (x^2 - y^2) \cdot (x^2 + y^2) \cdot (x^4 + y^4) \cdot (x^8 + y^8) \\
&= (x^4 - y^4) \cdot (x^4 + y^4) \cdot (x^8 + y^8) \\
&= (x^8 - y^8) \cdot (x^8 + y^8) \\
&= x^{16} - y^{16}
\end{aligned}$$

From here

$$x^{16} = x^{16} - y^{16}$$
$$x^{16} - x^{16} + y^{16} = 0$$
$$y^{16} = 0$$
$$y = 0$$

Since y is positive, then we obtained a contradiction.

Problem 3

Real numbers a, b, c satisfy the equations

$$a + b - c - d = 0$$
$$a^2 + b^2 - c^2 - d^2 = 0$$

Prove that they also satisfy the equation

$$a^n + b^n - c^n - d^n = 0$$

for all $n \geq 3$.

Solution

Let us rewrite the equations as

$$a - c = d - b$$
$$a^2 - c^2 = d^2 - b^2$$

Let us now factor both sides of the second equation using the Difference of Squares Formula:

$$(a - c)(a + c) = (d - b)(d + b)$$

We will proceed by doing the following casework:

- If $a - c = 0$, then $a = c$ and from the first equation $d - b = 0$, which implies that and $d = b$. From here the equality becomes

$$a^n + b^n - c^n - d^n = c^n + b^n - c^n - b^n = 0$$

  and holds as desired.

- If $d - b = 0$, then $d = b$ and from the first equation $a - c = 0$, which implies that and $a = c$. From here the equality becomes

$$a^n + b^n - c^n - d^n = c^n + b^n - c^n - b^n = 0$$

  and holds as desired.

- If $a - c \neq 0$ and $d - b \neq 0$, then we have

$$a + c = d + b$$

  Now adding and subtracting it with the equation $a - c = d - b$ we have $a = d$ and $b = c$. From here the equality becomes

$$a^n + b^n - c^n - d^n = d^n + c^n - c^n - d^n = 0$$

  and holds as desired.

CHAPTER 3

Square of a Sum and Square of a Difference Formulas

Many math olympiad problems rely on the **Square of a Sum and Square of a Difference Formulas**. These identities are very useful when working with expressions that involve squares.

Square of a Sum and Square of a Difference Formulas

1. **Square of a Sum Formula**:

$$(a + b)^2 = a^2 + 2ab + b^2$$

2. **Square of a Difference Formula**:

$$(a - b)^2 = a^2 - 2ab + b^2$$

Let us consider several problems.

Problem 1

Positive real numbers satisfy the equation

$$(a + c)^2 + (b - c)^2 = (a - c)^2 + (b + c)^2$$

Prove that $a = b$.

Solution

Let us rewrite the equation using the Square of a Sum Formula:

$$(a + c)^2 + (b - c)^2 = (a - c)^2 + (b + c)^2$$
$$a^2 + 2ac + c^2 + b^2 - 2bc + c^2 = a^2 - 2ac + c^2 + b^2 + 2bc + c^2$$
$$\cancel{a^2} + 2ac + \cancel{c^2} + \cancel{b^2} - 2bc + \cancel{c^2} = \cancel{a^2} - 2ac + \cancel{c^2} + \cancel{b^2} + 2bc + \cancel{c^2}$$
$$2ac - 2bc = -2ac + 2bc$$
$$4ac - 4bc = 0$$
$$4c(a - b) = 0$$

Since c is positive, then $a - b = 0$, or equivalently $a = b$, as desired.

Problem 2

Solve the equation in real numbers

$$(x + y + z)^2 = (x + y)^2 + (y + z)^2 + (z + x)^2$$

Solution

Answer: $x = y = z = 0$.

Let A denote the left-hand side and B denote the right-hand side of the equation.

Notice that we can rewrite A using the following identity[1]:

$$A = (x + y + z)^2$$
$$= x^2 + y^2 + z^2 + 2xy + 2yz + 2zx$$

[1] You can find more information about this topic in Chapter 9 "Advanced Identities".

Now we can rewrite B using the Square of a Sum Formula as follows:

$$B = (x+y)^2 + (y+z)^2 + (z+x)^2$$
$$= \left(x^2 + 2xy + y^2\right) + \left(y^2 + 2yz + z^2\right) + \left(z^2 + 2zx + x^2\right)$$
$$= 2x^2 + 2y^2 + 2z^2 + 2xy + 2yz + 2zx$$

The equation $A = B$ is now equivalent to

$$x^2 + y^2 + z^2 + 2xy + 2yz + 2zx = 2x^2 + 2y^2 + 2z^2 + 2xy + 2yz + 2zx$$
$$0 = x^2 + y^2 + z^2$$

Since each of the terms x^2, y^2 and z^2 is nonnegative, then their sum is equal to zero if and only if each of them is equal to zero. From here $x = y = z = 0$, as desired.

Problem 2

Let a and b be the legs, and c be the hypotenuse of some right triangle. Show that the equation

$$(x-a)^2 + (x-b)^2 = (x-c)^2$$

has no negative solutions for x.

Solution

Let us assume that there exists a negative solution x of the given equation.

Since a and b are the legs, and c is the hypotenuse of some right triangle, then by Pythagorean Theorem

$$a^2 + b^2 = c^2$$

or equivalently

$$a^2 + b^2 - c^2 = 0$$

Let us now apply the Square of a Difference Formula in the initial equation and rewrite it as follows:

$$(x-a)^2 + (x-b)^2 = (x-c)^2$$
$$x^2 - 2xa + a^2 + x^2 - 2xb + b^2 = x^2 - 2xc + c^2$$
$$x^2 - 2xa + a^2 + x^2 - 2xb + b^2 - x^2 + 2xc - c^2 = 0$$
$$x^2 - 2x(a+b-c) + \left(a^2 + b^2 - c^2\right) = 0$$

Taking into account that $a^2 + b^2 - c^2 = 0$ we have

$$x^2 - 2x(a+b-c) + \left(a^2 + b^2 - c^2\right) = 0$$
$$x^2 - 2x(a+b-c) = 0$$
$$x\left(x - 2(a+b-c)\right) = 0$$

From here $x = 0$ or $x - 2(a + b - c) = 0$. Since x is negative, then the first equality does not hold. From the second equality we have

$$x = 2(a + b - c)$$

However, by the triangle inequality we have

$$a + b > c$$

and therefore

$$x = 2(a + b - c) > 2(c - c) = 2(0) = 0$$

and therefore, x is positive. We obtained a contradiction.

CHAPTER 4

Completing the Square

Many math olympiad algebra problems are based on the procedure called **completing the square**. It relies on the **Inverted Square of a Sum and Square of a Difference Formulas** and is used to rewrite parts of the expression in square forms. Having squares is very useful to establish the nonnegativity of the expressions and obtain useful factorizations.

Inverted Square of a Sum and Square of a Difference Formulas

1. **Inverted Square of a Sum Formula**:

$$a^2 + 2ab + b^2 = (a + b)^2$$

2. **Inverted Square of a Difference Formula**:

$$a^2 - 2ab + b^2 = (a - b)^2$$

Let us consider several problems.

Problem 1

Real numbers a, b, c and d satisfy the equation

$$a^2 + b^2 + c^2 + d^2 = a(b + c + d)$$

Prove that $a + b + c + d = 0$.

Solution

First, let us rewrite the equation as

$$a^2 + b^2 + c^2 + d^2 - ab - ac - ad = 0$$

Let us now multiply this equation by 4 and complete the squares:

$$4a^2 + 4b^2 + 4c^2 + 4d^2 - 4ab - 4ac - 4ad = 0$$
$$a^2 + \left(a^2 - 4ab + 4b^2\right) + \left(a^2 - 4ac + 4c^2\right) + \left(a^2 - 4ad + 4d^2\right) = 0$$
$$a^2 + (a - 2b)^2 + (a - 2c)^2 + (a - 2d)^2 = 0$$

This immediately implies that $a = 0$, and, consequently, $b = 0$, $c = 0$ and $d = 0$. From here $a + b + c + d = 0$ as desired.

Problem 2

Solve the equation in real numbers

$$16^x + 25^x + 1 = 4^x + 5^x + 20^x$$

Solution

Let us rewrite the equation as

$$16^x + 25^x + 1 - 4^x - 5^x - 20^x = 0$$

Let us make the substitutions $m = 4^x$ and $n = 5^x$. We will now rewrite the equation in new variables, multiply it by 2 and complete the squares:

$$m^2 + n^2 + 1 - m - n - mn = 0$$
$$2m^2 + 2n^2 + 2 - 2m - 2n - 2mn = 0$$
$$\left(m^2 - 2m + 1\right) + \left(n^2 - 2n + 1\right) + \left(m^2 - 2mn + n^2\right) = 0$$
$$(m - 1)^2 + (n - 1)^2 + (m - n)^2 = 0$$

From here we have that $m = 1$ and $n = 1$. This implies that $x = 0$ is the only solution of the initial equation.[1]

Problem 3

Given positive real numbers a, b, c, d, such that

$$w^4 + x^4 + \frac{1}{y^4} + \frac{1}{z^4} - \frac{4wx}{yz} = 0$$

Prove that $w + y = x + z$.

Solution

Let us make the substitutions $a = w$, $b = x$, $c = \frac{1}{y}$, $d = \frac{1}{z}$. Then the equation becomes

$$a^4 + b^4 + c^4 + d^4 - 4abcd = 0$$

Let us complete the squares as follows:

$$a^4 + b^4 + c^4 + d^4 - 4abcd = 0$$
$$\left(a^4 - 2a^2b^2 + b^4\right) + \left(c^4 - 2c^2d^2 + d^4\right) + \left(2a^2b^2 - 4abcd + 2c^2d^2\right) = 0$$
$$\left(a^2 - b^2\right)^2 + \left(c^2 - d^2\right)^2 + 2\left(ab - cd\right)^2 = 0$$

From here we have the following system of equations:

$$\begin{cases} a^2 - b^2 = 0 \\ c^2 - d^2 = 0 \\ ab - cd = 0 \end{cases}$$

Since a, b, c, d are positive, then from the first two equations we have that $a = b$ and $c = d$. From the third equation we have that $a^2 - c^2 = 0$, which implies that $a = c$. From here we have that $a = b = c = d$. Therefore

$$w + y = a + \frac{1}{c} = b + \frac{1}{d} = x + z$$

which is what needed to be proven.[2]

---

[1] We applied a different approach to a similar problem in Chapter 35 "Increasing and Decreasing Functions".

[2] The same conclusion can be obtained by applying the AM-GM Inequality discussed in detail in Chapters 19–22.

CHAPTER 5

Sum of Cubes and Difference of Cubes Formulas

Many math olympiad problems use the **Sum of Cubes and Difference of Cubes Formulas**. These identities are very useful when working with expressions that involve cubes of the variables.

Sum of Cubes and Difference of Cubes Formulas

1. **Sum of Cubes Formula**:
$$a^3 + b^3 = (a + b)\left(a^2 - ab + b^2\right)$$

2. **Difference of Cubes Formula**:
$$a^3 - b^3 = (a - b)\left(a^2 + ab + b^2\right)$$

Let us consider several problems.

Problem 1

Nonzero numbers a and b satisfy the equalities

$$b + 1 = a^2 + \frac{1}{a^2}$$

$$b^2 + 1 = a^3 + \frac{1}{a^3}$$

Prove that

$$b + \frac{1}{b} = a + \frac{1}{a}$$

Solution

Let us consider the second equation and factor its right-hand side using the Sum of Cubes Formula:

$$b^2 + 1 = a^3 + \frac{1}{a^3}$$

$$b^2 + 1 = \left(a + \frac{1}{a}\right)\left(a^2 + \frac{1}{a^2} - 1\right)$$

Taking into account the conditions of the problem we have

$$b^2 + 1 = \left(a + \frac{1}{a}\right)\left(a^2 + \frac{1}{a^2} - 1\right)$$

$$b^2 + 1 = \left(a + \frac{1}{a}\right)(b + 1 - 1)$$

$$b^2 + 1 = \left(a + \frac{1}{a}\right)b$$

$$\frac{b^2 + 1}{b} = a + \frac{1}{a}$$

$$b + \frac{1}{b} = a + \frac{1}{a}$$

which is what needed to be proven.

Problem 2

Real numbers a, b and c satisfy the equalities

$$a - b = c$$
$$a^3 - b^3 = c^3$$

Prove that at least one of the numbers is zero.

Solution

Let us start by squaring the first equation:

$$a - b = c$$
$$(a - b)^2 = c^2$$
$$a^2 - 2ab + b^2 = c^2$$
$$a^2 + b^2 = c^2 + 2ab$$

Let us now factor the second equation using the Difference of Cubes Formula, and substitute the values of $a - b = c$ and $a^2 + b^2 = c^2 + 2ab$:

$$a^3 - b^3 = c^3$$
$$(a - b)\left(a^2 + ab + b^2\right) = c^3$$
$$c\left(c^2 + 3ab\right) = c^3$$
$$c^3 + 3abc = c^3$$
$$3abc = 0$$

From here at least one of the numbers a, b or c is zero, as desired.[1]

Problem 3

Given the distinct real numbers a, b, c, such that $abc = 1$ and

$$a^3 + \frac{1}{b} = b^3 + \frac{1}{c} = c^3 + \frac{1}{a}$$

Prove that
$$\left(a^2 + ab + b^2\right)\left(b^2 + bc + c^2\right)\left(c^2 + ca + a^2\right) = 1$$

[1] We solved this problem in a different way in Chapter 6 "Cube of a Sum and Cube of a Difference Formulas".

Solution

Let us put
$$X = \left(a^2 + ab + b^2\right)\left(b^2 + bc + c^2\right)\left(c^2 + ca + a^2\right)$$

Let us consider the first equality and apply the Difference of Cubes Formula:

$$a^3 + \frac{1}{b} = b^3 + \frac{1}{c}$$

$$a^3 - b^3 = \frac{1}{c} - \frac{1}{b}$$

$$(a - b)\left(a^2 + ab + b^2\right) = \frac{b}{bc} - \frac{c}{bc}$$

$$(a - b)\left(a^2 + ab + b^2\right) = \frac{b - c}{bc}$$

Similarly, we can show that

$$(b - c)\left(b^2 + bc + c^2\right) = \frac{c - a}{ca}$$

$$(c - a)\left(c^2 + ca + a^2\right) = \frac{a - b}{ab}$$

by cyclically permuting the variables.

Taking into account that the numbers a, b, c are distinct and $abc = 1$, we will proceed by multiplying these equations:

$$(a - b) \cdot (b - c) \cdot (c - a) \cdot X = \frac{b - c}{bc} \cdot \frac{c - a}{ca} \cdot \frac{a - b}{ab}$$

$$(a - b) \cdot (b - c) \cdot (c - a) \cdot X = \frac{(a - b) \cdot (b - c) \cdot (c - a)}{(abc)^2}$$

$$(a - b) \cdot (b - c) \cdot (c - a) \cdot X = (a - b) \cdot (b - c) \cdot (c - a)$$

$$X = 1$$

as desired.

CHAPTER 6

Cube of a Sum and Cube of a Difference Formulas

Many math olympiad problems rely on the **Cube of a Sum and Cube of a Difference Formulas**. These identities are very useful when working with expressions that involve cubes.

Cube of a Sum and Cube of a Difference Formulas

1. **Cube of a Sum Formula**:

$$(a + b)^3 = a^3 + 3a^2b + 3ab^2 + b^3$$

2. **Cube of a Difference Formula**:

$$(a - b)^3 = a^3 - 3a^2b + 3ab^2 - b^3$$

Let us consider several problems.

Problem 1

Prove that there do not exist distinct positive real numbers a, b, x that satisfy the equations

$$(a + x)^3 = (b + x)^3 + 1$$
$$a^3 = b^3 + 1$$

Solution

Let us assume that such numbers exist.

Let us subtract the equations, apply the Cube of a Sum Formula and use the Factorization by Grouping[1]:

$$(a + x)^3 - a^3 = (b + x)^3 + 1 - b^3 - 1$$
$$a^3 + 3a^2x + 3ax^2 + x^3 - a^3 = b^3 + 3b^2x + 3bx^2 + x^3 + 1 - b^3 - 1$$
$$3a^2x + 3ax^2 = 3b^2x + 3bx^2$$
$$3a^2x + 3ax^2 - 3b^2x - 3bx^2 = 0$$
$$3x\left(a^2 + ax - b^2 - bx\right) = 0$$
$$3x\left(\left(a^2 - b^2\right) + (ax - bx)\right) = 0$$
$$3x\left((a - b)(a + b) + x(a - b)\right) = 0$$
$$3x(a - b)(a + b + x) = 0$$

Notice that since a, b and x are positive, then $a + b + x$ is also positive.

This implies that from the last equation we have $a - b = 0$. From where $a = b$, which leads to a contradiction.

Problem 2

Real numbers a, b and c satisfy the equalities

$$a - b = c$$
$$a^3 - b^3 = c^3$$

Prove that at least one of the numbers is zero.

---

[1] You can find more information about this topic in Chapter 7 "Factorization by Grouping".

Solution

Notice that from the second equation we have

$$a^3 - b^3 = c^3$$
$$a^3 - b^3 - c^3 = 0$$

We will raise both sides of the first equation to the power 3 and apply the Cube of a Difference Formula:

$$a - b = c$$
$$(a - b)^3 = c^3$$
$$a^3 - 3a^2b + 3ab^2 - b^3 = c^3$$
$$a^3 - b^3 - c^3 = 3a^2b - 3ab^2$$

This implies that

$$3a^2b - 3ab^2 = 0$$

Taking into account that $a - b = c$ we have

$$3a^2b - 3ab^2 = 0$$
$$3ab(a - b) = 0$$
$$3abc = 0$$

and at least one of the numbers a, b or c is zero, as desired.[2]

Problem 3

Given $a + b + c + d = 0$, prove that

$$a^3 + b^3 + c^3 + d^3 = 3(abc + bcd + cda + dab)$$

Solution

Start by noticing that from the equality $a + b + c + d = 0$ we have

$$a + b = -c - d$$
$$c + d = -a - b$$

---

[2] We solved this problem in a different way in Chapter 5 "Sum of Cubes and Difference of Cubes Formulas".

We will raise both sides of the first equation to the power 3 and apply the Cube of a Sum Formula:

$$(a + b)^3 = -(c + d)^3$$

$$a^3 + 3a^2b + 3ab^2 + b^3 = -c^3 - 3c^2d - 3cd^2 - d^3$$

$$a^3 + b^3 + c^3 + d^3 = -3a^2b - 3ab^2 - 3c^2d - 3cd^2$$

$$a^3 + b^3 + c^3 + d^3 = 3ab(-a - b) + 3cd(-c - d)$$

$$a^3 + b^3 + c^3 + d^3 = 3ab(c + d) + 3cd(a + b)$$

$$a^3 + b^3 + c^3 + d^3 = 3(abc + bcd + cda + dab)$$

which is what needed to be proven.

CHAPTER 7

Factorization by Grouping

Typical Steps of Factorization by Grouping

Below we will obtain the factorization of the expression

$$xy + ax + by + ab$$

1. **Grouping**. Group the first two terms and the last two terms using parenthesis:
$$(xy + ax) + (by + ab)$$

2. **Common Factors**. Factor out the common factors in each parenthesis:
$$x(y + a) + b(y + a)$$

3. **Common Parenthesis**. Factor out the common parenthesis:
$$(y + a)(x + b)$$

Many math olympiad problems use the **Factorization by Grouping**. This factorization technique relies on grouping the terms in a strategic way and then factoring out common factors.

Let us now consider several problems.

Problem 1

Given two distinct nonzero real numbers a and b that satisfy the equation

$$a + \frac{1}{a} = b + \frac{1}{b}$$

Prove that one of the numbers is a reciprocal of another.

Solution

Notice that it will enough to prove that $ab = 1$.

Let us multiply the given equation by ab and rewrite it as follows:

$$a + \frac{1}{a} = b + \frac{1}{b}$$

$$ab \cdot \left(a + \frac{1}{a} \right) = ab \cdot \left(b + \frac{1}{b} \right)$$

$$a^2 b + b = ab^2 + a$$

$$a^2 b + b - ab^2 - a = 0$$

Let us now factor the expression on the left-hand side of the equation using the Factorization by Grouping:

$$a^2 b + b - ab^2 - a = 0$$

$$\left(a^2 b - ab^2 \right) - (a - b) = 0$$

$$ab(a - b) - (a - b) = 0$$

$$(a - b)(ab - 1) = 0$$

Since the numbers are distinct then from the last equation we have that $ab - 1 = 0$. This implies that $ab = 1$, and one of the numbers is a reciprocal of another, as desired.

Problem 2

Given integers x, y and z, such that

$$xy + yz + zx = 1$$

Prove that the number R defined as

$$R = \left(1 + x^2\right)\left(1 + y^2\right)\left(1 + z^2\right)$$

is a perfect square.

Solution

Even though this problem involves integer numbers, its solution is very algebraic.

Let us consider the quantity $1 + x^2$. Taking into account that $xy + yz + zx = 1$, let us factor it using the Factorization by Grouping:

$$
\begin{aligned}
1 + x^2 &= (xy + yz + zx) + x^2 \\
&= (xy + yz) + \left(x^2 + zx\right) \\
&= y\,(x + z) + x\,(x + z) \\
&= (x + z)(x + y)
\end{aligned}
$$

Similarly, we have

$$1 + y^2 = (y + x)(y + z)$$

and

$$1 + z^2 = (z + x)(z + y)$$

Therefore

$$
\begin{aligned}
R &= \left(1 + x^2\right)\left(1 + y^2\right)\left(1 + z^2\right) \\
&= (x + z)(x + y)(y + x)(y + z)(y + x)(y + z) \\
&= (x + y)^2(y + z)^2(z + x)^2 \\
&= \left((x + y)(y + z)(z + x)\right)^2
\end{aligned}
$$

as desired.

Problem 3

Given that

$$\frac{a}{b + c} = \frac{b}{c + a} = \frac{c}{a + b}$$

Find all possible values of the expression

$$S = \frac{a+b}{c} + \frac{b+c}{a} + \frac{c+a}{b}$$

Solution

Answer: $S = -3$ or $S = 6$.

Let consider the first equality. Let us cross-multiply, pass all terms to the left-hand side and factor it using the Factorization by Grouping:

$$a(a+c) = b(b+c)$$
$$a^2 + ac = b^2 + bc$$
$$a^2 + ac - b^2 - bc = 0$$
$$\left(a^2 - b^2\right) + (ac - bc) = 0$$
$$(a-b)(a+b) + c(a-b) = 0$$
$$(a-b)(a+b+c) = 0$$

Let consider the second equality. Let us cross-multiply, pass all terms to the left-hand side and factor it using the Factorization by Grouping:

$$b(a+b) = c(a+c)$$
$$ab + b^2 = ac + c^2$$
$$ab + b^2 - ac - c^2 = 0$$
$$\left(b^2 - c^2\right) + (ab - ac) = 0$$
$$(b-c)(b+c) + a(b-c) = 0$$
$$(b-c)(a+b+c) = 0$$

We will proceed by doing the following casework:

- If $a + b + c = 0$, then $a + b = -c$, $b + c = -a$, $a + c = -b$ and

$$S = \frac{-c}{c} + \frac{-a}{a} + \frac{-b}{b} = -1 - 1 - 1 = -3$$

- If $a + b + c \neq 0$, then $a - b = 0$ and $b - c = 0$, which implies that $a = b = c$ and

$$S = \frac{2c}{c} + \frac{2a}{a} + \frac{2b}{b} = 2 + 2 + 2 = 6$$

CHAPTER 8

Simon's Favorite Factoring Trick

Many math olympiad problems rely on **Simon's Favorite Factoring Trick**. This is a factorization technique that is based on factorization by grouping and allows for partial factoring of certain expressions. It involves adding and subtracting a carefully chosen constant term to create a factorable expression.

Typical Steps of Simon's Favorite Factoring Trick

Below we will obtain the partial factorization of the expression

$$xy + ax + by + c$$

1. **Grouping**. Group some two terms using parenthesis:

$$(xy + ax) + by + c$$

---

2. **Common Factor**. Factor out the common factor in the parenthesis:

$$x(y+a) + by + c$$

3. **Adding and Subtracting**. Add and subtract ab:

$$x(y+a) + (by + ab) - ab + c$$

4. **Common Factor**. Factor out the common factor in the second parenthesis:

$$x(y+a) + b(y+a) - ab + c$$

5. **Common Parenthesis**. Factor out the common parenthesis:

$$(y+a)(x+b) - ab + c$$

Let us now consider several problems.

Problem 1

Find all integers a and b that satisfy the equation

$$ab + 2a = b + 1$$

Solution

Answer: $(a, b) \in \{(0, -1), (2, -3)\}$.

Even though this problem involves integer numbers, its solution is very algebraic.

Let us pass b to the left-hand side and use the Simon's Favorite Factoring Trick:

$$ab + 2a = b + 1$$
$$ab + 2a - b = 1$$
$$a(b+2) - b = -1$$
$$a(b+2) - (b+2) + 2 = 1$$
$$a(b+2) - (b+2) = -1$$
$$(b+2)(a-1) = -1$$

Notice that the product of two integers is -1 only when the numbers are -1 and 1, or 1 and -1.

We will proceed by doing the following casework:

- If $b + 2 = 1$ and $a - 1 = -1$, then $b = -1$ and $a = 0$.
- If $b + 2 = -1$ and $a - 1 = 1$, then $b = -3$ and $a = 2$.

Problem 2

Given real nonzero numbers x and y, such that

$$\frac{1}{x} + \frac{1}{y} + 1 = 0$$

Prove that for all positive integer n:

$$(x + 1)^{2n} + (y + 1)^{2n} \geq 2$$

Solution

Let us rewrite the given equality as follows:

$$\frac{1}{x} + \frac{1}{y} + 1 = 0$$

$$\frac{y}{xy} + \frac{x}{xy} + \frac{xy}{xy} = 0$$

$$\frac{xy + x + y}{xy} = 0$$

This implies that $xy + x + y = 0$. Let us use the Simon's Favorite Factoring Trick to factor the left-hand side of this equality:

$$xy + x + y = 0$$
$$x(y + 1) + y = 0$$
$$x(y + 1) + (y + 1) - 1 = 0$$
$$(y + 1)(x + 1) = 1$$

From here using the AM-GM Inequality[1] we have

$$(x + 1)^{2n} + (y + 1)^{2n} \geq 2\sqrt{(x + 1)^{2n}(y + 1)^{2n}}$$

$$= 2\sqrt{((x + 1)(y + 1))^{2n}}$$

$$= 2$$

as desired.

[1] You can find more information about this topic in Chapters 19–22.

Problem 3

Nonzero real numbers a, b, c satisfy the equality

$$\frac{1}{a} + \frac{1}{b} = \frac{1}{c}$$

Prove that if $c < a$, then $a + b$ is positive.

Solution

Let us rewrite the given equality and cross-multiply:

$$\frac{1}{a} + \frac{1}{b} = \frac{1}{c}$$
$$\frac{b}{ab} + \frac{a}{ab} = \frac{1}{c}$$
$$\frac{a+b}{ab} = \frac{1}{c}$$
$$c(a+b) = ab$$

Let us pass al terms to the left-hand side and use the Simon's Favorite Factoring Trick:

$$c(a+b) = ab$$
$$c(a+b) - ab = 0$$
$$c(a+b) - \left(a^2 + ab\right) + a^2 = 0$$
$$c(a+b) - a(a+b) + a^2 = 0$$
$$(a+b)(c-a) + a^2 = 0$$
$$(a+b)(c-a) = -a^2$$

Notice that since a is nonzero, then the right-hand side is negative, and so should be left-hand side. Since $c < a$, then $c - a$ is negative. This implies that $a + b$ is positive, as desired.

CHAPTER 9

Advanced Identities

Many algebra problems rely on **Advanced Identities**, which are very useful when manipulating expressions that resemble some specific components of these formulas.

Advanced Identities

1. $(a+b+c)^2 = a^2 + b^2 + c^2 + 2(ab + bc + ca)$

2. $(a^2 + b^2)(c^2 + d^2) = (ac + bd)^2 + (ad - bc)^2$

3. $(a + b + c)(bc + ca + ab) = (a + b)(b + c)(c + a) + abc$

4. $a^3 + b^3 + c^3 - 3abc = (a + b + c)(a^2 + b^2 + c^2 - ab - bc - ca)$

5. $(a + b + c)^3 - a^3 - b^3 - c^3 = 3(b + c)(c + a)(a + b)$

6. $a^4 + b^4 + c^4 = (a^2 + b^2 + c^2)^2 - 2(a^2b^2 + b^2c^2 + c^2a^2)$

7. $a^2(c - b) + b^2(a - c) + c^2(b - a) = (a - b)(b - c)(c - a)$

Let us consider several problems.

Problem 1

Given the nonzero real numbers a, b and c, such that

$$\frac{1}{a+b+c} = \frac{1}{a} + \frac{1}{b} + \frac{1}{c}$$

Show that one of the numbers a, b, c is an additive inverse of another.

Solution

Let us rewrite the equation as follows:

$$\frac{1}{a+b+c} = \frac{1}{a} + \frac{1}{b} + \frac{1}{c}$$

$$\frac{1}{a+b+c} = \frac{ab+bc+ca}{abc}$$

After the cross-multiplication we have

$$abc = (a+b+c)\,(ab+bc+ca)$$

Taking into account the fundamental identity #3 we have

$$abc = (a+b)(b+c)(c+a) + abc$$

which implies that

$$(a+b)(b+c)(c+a) = 0$$

From here one of the parenthesis should equal zero. Therefore, one of the numbers a, b, c is an additive inverse of another, as desired.

Problem 2

The equality
$$(a+b+c)^3 = a^3 + b^3 + c^3$$

holds for some nonzero real numbers a, b and c. Find all positive integer numbers n, such that
$$(a+b+c)^n = a^n + b^n + c^n$$

Solution

From the fundamental identity #5 we immediately have that

$$(a+b)(b+c)(c+a) = 0$$

Without loss of generality let us assume that $a+b=0$, which implies that $b=-a$. We will proceed by doing the following casework:

- If n is odd, then the equation is equivalent to

$$a^n + (-a)^n + c^n = (a - a + c)^n$$
$$c^n = c^n$$

  which obviously holds.

- If n is even, then the equation is equivalent to

$$a^n + (-a)^n + c^n = (a - a + c)^n$$
$$2a^n + c^n = c^n$$
$$2a^n = 0$$

  which contradicts to the fact that a is nonzero.

We conclude that n is any odd positive integer.

Problem 3

Integer numbers satisfy the condition $a+b+c=0$. Prove that

$$2a^4 + 2b^4 + 2c^4$$

is a perfect square.

Solution

Even though this problem involves integer numbers, its solution is very algebraic.

From the fundamental identity #1 we have

$$(a+b+c)^2 = (0)^2$$
$$a^2 + b^2 + c^2 + 2ab + 2bc + 2ac = 0$$
$$a^2 + b^2 + c^2 = -2ab - 2bc - 2ac$$

From the fundamental identity #6 we have

$$a^4 + b^4 + c^4 = \left(a^2 + b^2 + c^2\right)^2 - 2\left(a^2b^2 + b^2c^2 + a^2c^2\right)$$
$$a^4 + b^4 + c^4 = \left(-2ab - 2bc - 2ac\right)^2 - 2\left(a^2b^2 + b^2c^2 + a^2c^2\right)$$
$$a^4 + b^4 + c^4 = 2\left(a^2b^2 + b^2c^2 + a^2c^2\right) + 8\left(a^2bc + ab^2c + abc^2\right)$$
$$2a^4 + 2b^4 + 2c^4 = 4\left(a^2b^2 + b^2c^2 + a^2c^2\right) + 16\left(a^2bc + ab^2c + abc^2\right)$$

Notice that since $a + b + c = 0$, then

$$a^2bc + ab^2c + abc^2 = abc(a + b + c) = 0$$

Therefore

$$2a^4 + 2b^4 + 2c^4 = 4\left(a^2b^2 + b^2c^2 + a^2c^2\right)$$
$$2a^4 + 2b^4 + 2c^4 = 4\left(\left(a^2b^2 + b^2c^2 + a^2c^2\right) + 2\left(a^2bc + ab^2c + abc^2\right)\right)$$
$$2a^4 + 2b^4 + 2c^4 = 4\left(\left(a^2b^2 + b^2c^2 + a^2c^2\right) + 2\left(ab \cdot ac + ab \cdot bc + bc \cdot ac\right)\right)$$
$$2a^4 + 2b^4 + 2c^4 = 4\left(ab + bc + ac\right)^2$$
$$2a^4 + 2b^4 + 2c^4 = \left(2ab + 2bc + 2ac\right)^2$$

which is what needed to be proven.[1]

---

[1] We solved this problem in a different way in Chapter 14 "Vieta's Formulas for Cubic Polynomial".

CHAPTER 10

Linear Equations in One Variable

Linear Equation

Given real numbers a and b. Equation of the form

$$ax + b = 0$$

is called **linear** in terms of the variable x.

In certain math olympiad problems, the task at hand involves working with a specific **linear equation**. These equations can often appear significantly more intricate than the one defined above. Successfully tackling these problems typically entails employing various algebraic manipulations, formulas and factorization techniques discussed in the previous chapters.

Let us consider several problems.

Problem 1

Given positive real numbers a, b, c, such that $a + b + c = 1$. Show that the equation

$$\frac{a^2 x}{b + c - 1} + \frac{b^2 x}{c + a - 1} + \frac{c^2 x}{a + b - 1} = 1$$

has a negative solution for x.

Solution

Taking into account that $a + b + c = 1$, we have

$$b + c - 1 = -a$$
$$c + a - 1 = -b$$
$$a + b - 1 = -c$$

Let us now rewrite the left-hand side of the equation as follows:

$$\frac{a^2 x}{b + c - 1} + \frac{b^2 x}{c + a - 1} + \frac{c^2 x}{a + b - 1} = 1$$
$$\frac{a^2 x}{-a} + \frac{b^2 x}{-b} + \frac{c^2 x}{-c} = 1$$
$$-ax - bx - cx = 1$$
$$-x(a + b + c) = 1$$
$$-x = 1$$
$$x = -1$$

From here we see that the given equation has a negative solution for x, as desired.

Problem 2

Given three positive numbers a, b, c. Solve the equation

$$\frac{x - a - b}{c} + \frac{x - b - c}{a} + \frac{x - c - a}{b} = 3$$

for x and simplify your answer.

Solution

Answer: $x = a + b + c$.

Let us start by splitting 3 as $3 = 1 + 1 + 1$, and pass it to the left-hand side of the equation:

$$\frac{x-a-b}{c} + \frac{x-b-c}{a} + \frac{x-c-a}{b} = 3$$

$$\frac{x-a-b}{c} + \frac{x-b-c}{a} + \frac{x-c-a}{b} = 1 + 1 + 1$$

$$\frac{x-a-b}{c} - 1 + \frac{x-b-c}{a} - 1 + \frac{x-c-a}{b} - 1 = 0$$

Let us now combine the fractions on the left-hand side of the equation:

$$\frac{x-a-b}{c} - 1 + \frac{x-b-c}{a} - 1 + \frac{x-c-a}{b} - 1 = 0$$

$$\frac{x-a-b}{c} - \frac{c}{c} + \frac{x-b-c}{a} - \frac{a}{a} + \frac{x-c-a}{b} - \frac{b}{b} = 0$$

$$\frac{x-a-b-c}{c} + \frac{x-a-b-c}{a} + \frac{x-a-b-c}{b} = 0$$

Notice that the left-hand side has a common factor[1] of $x - a - b - c$. Let us factor it out and rewrite the equation as follows:

$$\frac{x-a-b-c}{c} + \frac{x-a-b-c}{a} + \frac{x-a-b-c}{b} = 0$$

$$(x - a - b - c)\left(\frac{1}{a} + \frac{1}{b} + \frac{1}{c}\right) = 0$$

Since a, b and c are positive numbers, then so is $\frac{1}{a} + \frac{1}{b} + \frac{1}{c}$. From here

$$x - a - b - c = 0$$

or equivalently

$$x = a + b + c$$

as desired.

Problem 3

Given positive real numbers a, b, c, such that $abc = 1$. Solve the equation

$$\frac{ax}{ab+a+1} + \frac{bx}{bc+b+1} + \frac{cx}{ca+c+1} = -1$$

in real numbers for the variable x.

[1] You can find more information about this topic in Chapter 1 "Common Factor".

Solution

Let us multiply the numerator and the denominator of the first fraction by bc:

$$\frac{ax}{ab + a + 1} = \frac{abcx}{ab^2c + abc + bc}$$

and the numerator and the denominator of the third fraction by b:

$$\frac{cx}{ca + c + 1} = \frac{bcx}{abc + bc + b}$$

Taking into account that $abc = 1$, we can rewrite the initial equation as follows

$$\frac{ax}{ab + a + 1} + \frac{bx}{bc + b + 1} + \frac{cx}{ca + c + 1} = -1$$

$$\frac{abcx}{ab^2c + abc + bc} + \frac{bx}{bc + b + 1} + \frac{bcx}{abc + bc + b} = -1$$

$$\frac{x}{b + 1 + bc} + \frac{bx}{bc + b + 1} + \frac{bcx}{1 + bc + b} = -1$$

$$\frac{x + bx + bcx}{1 + b + bc} = -1$$

$$\frac{x(1 + b + bc)}{1 + b + bc} = -1$$

$$x = -1$$

We conclude that the solution of the given equation is $x = -1$.

CHAPTER 11

Polynomial Equations in One Variable

Polynomial Equations in One Variable

Polynomial equations in one variable are the equations of the form:

$$a_n x^n + a_{n-1} x^{n-1} + \ldots + a_2 x^2 + a_1 x + a_0 = 0$$

where x is the variable and a_i are given coefficients with $a_n \neq 0$.

In this chapter we will solve several polynomial equations. It is customary to employ such techniques as guessing the solution, factorization, change of variables and completing the square in these types of problems.

Let us consider several problems.

Problem 1

Solve the equation in positive real numbers

$$x^4 + x^2 = 1 + 2x^3$$

Solution

Answer: $x = \frac{1+\sqrt{5}}{2}$.

Let us pass $2x^3$ to the left-hand side and complete the square:

$$x^4 + x^2 = 1 + 2x^3$$
$$x^4 - 2x^3 + x^2 = 1$$
$$\left(x^2 - x\right)^2 = 1$$

From here $x^2 - x = \pm 1$.

We will proceed by doing the following casework:

- If $x^2 - x = 1$, then $x^2 - x - 1 = 0$, and we can apply the Quadratic Formula:

$$x = \frac{-(-1) \pm \sqrt{(-1)^2 - 4(1)(-1)}}{2(1)}$$
$$= \frac{1 \pm \sqrt{1+4}}{2}$$
$$= \frac{1 \pm \sqrt{5}}{2}$$

  Since we are looking for positive values, then $x = \frac{1+\sqrt{5}}{2}$.

- If $x^2 - x = -1$, then $x^2 - x + 1 = 0$. The discriminant is equal to

$$D = b^2 - 4ac = (-1)^2 - 4(1)(1) = -3$$

  and there are no real solutions in this case.

We conclude that the equation has a unique positive solution $x = \frac{1+\sqrt{5}}{2}$.

Problem 2

Solve the equation in real numbers

$$x^4 + x + 1 = x^3 + 2x^2$$

Solution

Answer: $x = \pm 1$, $x = \frac{1 \pm \sqrt{5}}{2}$.

Let us pass all the terms to the left-hand side and use the Factorization by Grouping:

$$x^4 + x + 1 = x^3 + 2x^2$$

$$x^4 - x^3 - 2x^2 + x + 1 = 0$$

$$\left(x^4 - x^3\right) - \left(2x^2 - 2x\right) - (x - 1) = 0$$

$$x^3(x - 1) - 2x(x - 1) - (x - 1) = 0$$

$$(x - 1)\left(x^3 - 2x - 1\right) = 0$$

From here we have $x - 1 = 0$ or $x^3 - 2x - 1 = 0$. The first equation implies that $x = 1$. Let us factor the left-hand side of the second equation using the Factorization by Grouping:

$$x^3 - 2x - 1 = 0$$

$$\left(x^3 + x^2\right) - \left(x^2 + x\right) - (x + 1) = 0$$

$$x^2(x + 1) - x(x + 1) - (x + 1) = 0$$

$$(x + 1)\left(x^2 - x - 1\right) = 0$$

From here we have $x + 1 = 0$ or $x^2 - x - 1 = 0$. The first equation implies that $x = -1$. For the second equation we will us apply the Quadratic Formula[1]:

$$x = \frac{-(-1) \pm \sqrt{(-1)^2 - 4(1)(-1)}}{2(1)}$$

$$= \frac{1 \pm \sqrt{1 + 4}}{2}$$

$$= \frac{1 \pm \sqrt{5}}{2}$$

We conclude that $x = \pm 1$, $x = \frac{1 \pm \sqrt{5}}{2}$.

Problem 3

Solve the equation

$$x^4 + 5x^2 + 1 = 4x^3 + 4x$$

in real numbers.

Answer: $x = \frac{3 \pm \sqrt{5}}{2}$.

---

[1] You can find more information about this topic in Chapter 12 "Properties of Quadratic Functions".

Solution

Let us divide both sides of the original equation by x^2:

$$x^4 + 5x^2 + 1 = 4x^3 + 4x$$

$$\frac{x^4}{x^2} + \frac{5x^2}{x^2} + \frac{1}{x^2} = \frac{4x^3}{x^2} + \frac{4x}{x^2}$$

$$x^2 + 5 + \frac{1}{x^2} = 4x + \frac{4}{x}$$

$$x^2 + 5 + \frac{1}{x^2} = 4\left(x + \frac{1}{x}\right)$$

$$x^2 + \frac{1}{x^2} = 4\left(x + \frac{1}{x}\right) - 5$$

Let us put $k = x + \frac{1}{x}$, square it and apply the Square of Sum Formula:

$$k = x + \frac{1}{x}$$

$$k^2 = \left(x + \frac{1}{x}\right)^2$$

$$k^2 = x^2 + 2 + \frac{1}{x^2}$$

$$k^2 - 2 = x^2 + \frac{1}{x^2}$$

We can now rewrite the original equation in terms of k:

$$x^2 + \frac{1}{x^2} = 4\left(x + \frac{1}{x}\right) - 5$$

$$k^2 - 2 = 4k - 5$$

$$k^2 - 4k + 3 = 0$$

$$(k - 1)(k - 3) = 0$$

From here we have that $k = 1$ or $k = 3$. We will proceed by doing the following casework:

- If $k = 1$, then we have

$$k = x + \frac{1}{x}$$

$$1 = x + \frac{1}{x}$$

$$x = x^2 + 1$$

$$0 = x^2 - x + 1$$

The discriminant of the last equation is equal to

$$D = b^2 - 4ac = (-1)^2 - 4(1)(1) = -3$$

and there are no real solutions in this case.

- If $k = 3$, then we have

$$k = x + \frac{1}{x}$$

$$3 = x + \frac{1}{x}$$

$$3x = x^2 + 1$$

$$0 = x^2 - 3x + 1$$

Now we can apply the Quadratic Formula[2]:

$$x = \frac{-(-3) \pm \sqrt{(-3)^2 - 4(1)(1)}}{2(1)}$$

$$= \frac{3 \pm \sqrt{9 - 4}}{2}$$

$$= \frac{3 \pm \sqrt{5}}{2}$$

We conclude that the real solutions of the given equation are $x = \frac{3 \pm \sqrt{5}}{2}$.

---

[2] You can find more information about this topic in Chapter 12 "Properties of Quadratic Functions".

CHAPTER 12

Properties of Quadratic Functions

In many math olympiad problems, the properties of **quadratic functions** come into play as invaluable tools. The usual approach often involves considering a particular expression as quadratic with respect to one of the variables. This becomes particularly useful when we need to determine the number of solutions of the equations, proving inequalities and finding the extrema of the underlying expressions.[1]

Quadratic Function

Quadratic function is a function of the form

$$f(x) = ax^2 + bx + c$$

where a, b and c are real numbers and $a \neq 0$.

---

[1]You can find some additional properties of quadratic functions in Chapter 13 "Vieta's Formulas for Quadratic Polynomial".

Roots of Quadratic Function

The roots of the quadratic function $f(x) = ax^2 + bx + c$ are given by the formula

$$x = \frac{-b \pm \sqrt{b^2 - 4ac}}{2a}$$

Discriminant of Quadratic Function

The discriminant D of the quadratic function $f(x) = ax^2 + bx + c$ is defined as

$$D = b^2 - 4ac$$

Number of Roots of Quadratic Function

The number of real roots of the quadratic function $f(x) = ax^2 + bx + c$ is determined from the following:

- If $D > 0$, then the function $f(x)$ has exactly two distinct real roots.
- If $D = 0$, then the function $f(x)$ has exactly one real root.
- If $D < 0$, then the function $f(x)$ has no real roots.

Maximum and Minimum of Quadratic Function

The maximum and the minimum of the quadratic function $f(x) = ax^2 + bx + c$ are determined from the following:

- If $a > 0$, then the minimum of the function $f(x)$ is reached at $x = -\frac{b}{2a}$.
- If $a < 0$, then the maximum of the function $f(x)$ is reached at $x = -\frac{b}{2a}$.

Let us consider several problems.

Problem 1

Given nonzero real numbers a, b and c. Show that the equation

$$\left(x - \frac{a}{b}\right)\left(x - \frac{b}{c}\right) + \left(x - \frac{b}{c}\right)\left(x - \frac{c}{a}\right) + \left(x - \frac{c}{a}\right)\left(x - \frac{a}{b}\right) = 0$$

has at least one real solution.

Solution

Let the left-hand side of the equation be N. Let us expand the expression for N and rewrite it as follows:

$$N = \left(x - \frac{a}{b}\right)\left(x - \frac{b}{c}\right) + \left(x - \frac{b}{c}\right)\left(x - \frac{c}{a}\right) + \left(x - \frac{c}{a}\right)\left(x - \frac{a}{b}\right)$$

$$= x^2 - \left(\frac{a}{b} + \frac{b}{c}\right)x + \frac{a}{c} + x^2 - \left(\frac{b}{c} + \frac{c}{a}\right)x + \frac{b}{a} + x^2 - \left(\frac{c}{a} + \frac{a}{b}\right)x + \frac{c}{b}$$

$$= 3x^2 - 2\left(\frac{a}{b} + \frac{b}{c} + \frac{c}{a}\right)x + \left(\frac{b}{a} + \frac{c}{b} + \frac{a}{c}\right)$$

Let us consider the quadratic function

$$f(t) = 3t^2 - 2\left(\frac{a}{b} + \frac{b}{c} + \frac{c}{a}\right)t + \left(\frac{b}{a} + \frac{c}{b} + \frac{a}{c}\right)$$

and its discriminant

$$D = 4\left(\frac{a}{b} + \frac{b}{c} + \frac{c}{a}\right)^2 - 12\left(\frac{b}{a} + \frac{c}{b} + \frac{a}{c}\right)$$

We will prove that $D \geq 0$, which will automatically imply that the function $f(x)$ has at least one real root.

The inequality $D \geq 0$ is equivalent to

$$4\left(\frac{a}{b} + \frac{b}{c} + \frac{c}{a}\right)^2 - 12\left(\frac{b}{a} + \frac{c}{b} + \frac{a}{c}\right) \geq 0$$

$$\left(\frac{a}{b} + \frac{b}{c} + \frac{c}{a}\right)^2 - 3\left(\frac{b}{a} + \frac{c}{b} + \frac{a}{c}\right) \geq 0$$

$$\frac{a^2}{b^2} + \frac{b^2}{c^2} + \frac{c^2}{a^2} + 2\left(\frac{a}{b}\cdot\frac{b}{c} + \frac{b}{c}\cdot\frac{c}{a} + \frac{c}{a}\cdot\frac{a}{b}\right) - 3\left(\frac{b}{a} + \frac{c}{b} + \frac{a}{c}\right) \geq 0$$

$$\frac{a^2}{b^2} + \frac{b^2}{c^2} + \frac{c^2}{a^2} + 2\left(\frac{b}{a} + \frac{c}{b} + \frac{a}{c}\right) - 3\left(\frac{b}{a} + \frac{c}{b} + \frac{a}{c}\right) \geq 0$$

$$\frac{a^2}{b^2} + \frac{b^2}{c^2} + \frac{c^2}{a^2} - \frac{b}{a} - \frac{c}{b} - \frac{a}{c} \geq 0$$

The last inequality follows from the Rearrangement Inequality[2] applied to the numbers $\left(\frac{a}{b}, \frac{b}{c}, \frac{c}{a}\right)$ and $\left(\frac{a}{b}, \frac{b}{c}, \frac{c}{a}\right)$:

$$\frac{a^2}{b^2} + \frac{b^2}{c^2} + \frac{c^2}{a^2} \geq \frac{a}{b} \cdot \frac{b}{c} + \frac{b}{c} \cdot \frac{c}{a} + \frac{c}{a} \cdot \frac{a}{b} = \frac{b}{a} + \frac{c}{b} + \frac{a}{c}$$

Problem 2

Prove that for the real numbers x, y and z the following inequality holds

$$x^2 \left(3y^2 + 3z^2 - 2yz\right) \geq yz \left(2xy + 2xz - yz\right)$$

Solution

Start by noticing that if $yz = 0$, then the inequality becomes

$$x^2(3y^2 + 3z^2) \geq 0$$

and obviously holds.

From now on we will assume that $yz \neq 0$. Let us show that

$$3y^2 + 3z^2 - 2yz > 0$$

Indeed, if $yz < 0$, then

$$3y^2 + 3z^2 - 2yz > 3y^2 + 3z^2 \geq 0$$

and the inequality holds.

If $yz > 0$, then

$$3y^2 + 3z^2 - 2yz > 3y^2 + 3z^2 - 6yz = 3(y - z)^2 \geq 0$$

and the inequality holds.

Now let us put $y + z = A$ and $yz = B$. Then

$$3y^2 + 3z^2 - 2yz = 3(y + z)^2 - 8yz = 3A^2 - 8B$$

and we have that

$$3A^2 - 8B > 0$$

The inequality of the problem now becomes

$$x^2 \left(3A^2 - 8B\right) \geq B(2xA - B)$$

---

[2]This inequality is discussed in detail in Chapter 30 "Rearrangement Inequality".

or equivalently

$$\left(3A^2 - 8B\right)x^2 - (2AB)\,x + B^2 \geq 0$$

Let us now consider the quadratic function

$$f(t) = \left(3A^2 - 8B\right)t^2 - (2AB)\,t + B^2$$

It will be enough to prove that $f(t) \geq 0$ for all $t \in \mathbb{R}$. Notice that

$$A^2 - 4B = (y+z)^2 - 4yz = y^2 - 2yz + z^2 = (y-z)^2 \geq 0$$

Therefore, the discriminant of the function $f(t)$ is

$$D = (2AB)^2 - 4\left(3A^2 - 8B\right)B^2 = -8B^2\left(A^2 - 4B\right) \leq 0$$

Since the leading coefficient of the function $f(t)$ is positive and its discriminant is nonpositive, then $f(t) \geq 0$ for all real numbers t, as desired.

Problem 3

Let a, b, c, d be distinct real numbers, such that $ac = bd$. Given that

$$\frac{a}{b} + \frac{b}{c} + \frac{c}{d} + \frac{d}{a} = 4$$

Find the maximum value of the expression

$$M = \frac{a}{c} + \frac{b}{d} + \frac{c}{a} + \frac{d}{b}$$

Solution

Let us make the following substitutions:

$$\frac{a}{b} = x$$

$$\frac{b}{c} = y$$

Notice that since a, b, c, d are distinct, then $x \neq 1$ and $y \neq 1$. Furthermore, we have

$$\frac{a}{c} = xy$$

$$\frac{c}{a} = \frac{1}{xy}$$

$$\frac{c}{d} = \frac{b}{a} = \frac{1}{x}$$

$$\frac{d}{a} = \frac{c}{b} = \frac{1}{y}$$

Also

$$\frac{b}{d} = \frac{a}{d} \cdot \frac{c}{d} = \frac{y}{x}$$

and

$$\frac{d}{b} = \frac{x}{y}$$

Therefore, the problem can be rewritten as follows: find the maximum of

$$M = xy + \frac{x}{y} + \frac{1}{xy} + \frac{y}{x}$$

given that

$$\left(x + \frac{1}{x}\right) + \left(y + \frac{1}{y}\right) = 4$$

Notice that from the last equality x and y cannot be both positive. Indeed, for positive x and y by the AM-GM Inequality[3]

$$4 = \left(x + \frac{1}{x}\right) + \left(y + \frac{1}{y}\right) \geq (2) + (2) = 4$$

However, this means that in the AM-GM Inequality the equality holds, which in turn implies that $x = y = 1$ and leads to a contradiction.

Therefore, at least one of x or y is negative. Without loss of generality let it be x.

Let us factor the expression for M:

$$M = xy + \frac{x}{y} + \frac{1}{xy} + \frac{y}{x} = x\left(y + \frac{1}{y}\right) + \frac{1}{x}\left(\frac{1}{y} + y\right) = \left(x + \frac{1}{x}\right)\left(y + \frac{1}{y}\right)$$

If we put $x + \frac{1}{x} = t$, then $y + \frac{1}{y} = 4 - t$. Then the expression for M becomes

$$M = t(4 - t) = -t^2 + 4t$$

Notice that since x is negative, then by the AM-GM Inequality applied to the numbers $-x$ and $\frac{1}{-x}$ we have

$$-x + \frac{1}{-x} \geq 2$$

which implies that

$$x + \frac{1}{x} \leq -2$$

and, therefore, $t \leq -2$.

Let us now consider the quadratic function $f(t) = -t^2 + 4t$. The maximum of the function $f(t)$ is located at $t = 2$, and the function is increasing on the interval

---

[3] The AM-GM Inequality is discussed in detail in Chapters 19–22.

53

$(-\infty, 2]$. However, since in our case $t \leq -2$, then the maximum of the function $f(t)$ on the interval $(-\infty, -2]$ is reached at $t = -2$, which in turn is reached for $x = -1$ and $a = -b$. From here we have

$$M = f(t) \leq f(-2) = -(-2)^2 + 4(-2) = -12$$

CHAPTER 13

Vieta's Formulas for Quadratic Polynomial

Vieta's Formulas for Quadratic Polynomial are a set of relations that establish connections between the coefficients of a quadratic polynomial and its roots. These formulas are particularly useful when we need to deal with products and sums of roots of quadratic polynomials.

Vieta's Formulas for Quadratic Polynomial

If x_1 and x_2 are the roots of the polynomial

$$f(x) = ax^2 + bx + c$$

where $a \neq 0$, then the following equalities hold

$$x_1 + x_2 = -\frac{b}{a}$$

$$x_1 \cdot x_2 = \frac{c}{a}$$

Let us consider several problems.

Problem 1

Find all tuples of real numbers (a, b, c, d), such that the equation $x^2 + ax + b = 0$ has solutions $x = c$ and $x = d$, and the equation $x^2 + cx + d = 0$ has solutions $x = a$ and $x = b$.

Solution

From the Vieta's Formulas we have

$$-a = c + d$$
$$-c = a + b$$
$$d = ab$$
$$b = cd$$

From the first two equations we have that

$$d = -a - c$$
$$b = -a - c$$

and thus $b = d$. Substituting $b = d$ into the third and fourth equations we have

$$d = ad$$
$$d = cd$$

or equivalently

$$d(a - 1) = 0$$
$$d(c - 1) = 0$$

We will proceed by doing the following casework:

- If $d = 0$, then $b = 0$ and $c = -a$. Therefore, the initial equations are $x^2 + ax = 0$ and $x^2 - ax = 0$, which satisfy the conditions of the problem. We conclude that the tuples of the form $(a, 0 - a, 0)$ are the solutions for all $a \in \mathbb{R}$.

- If $d \neq 0$, then $a = 1$ and $c = 1$. From here $d = -2$ and the initial equations are $x^2 + x - 2 = 0$ and $x^2 + x - 2 = 0$, which satisfy the conditions of the problem. We conclude that the tuple $(1, -2, 1, -2)$ is also a solution.

Problem 2

Line l intersects the parabola $y = x^2$ at the points A and B and the x-axis at the distinct point C. Let a, b, c be the x-coordinates of the points A, B and C respectfully. Prove that

$$\frac{1}{a} + \frac{1}{b} = \frac{1}{c}$$

Solution

Let m be the slope of the line l. Since the line l passes through the point $(c, 0)$, then its equation is

$$y = m(x - c)$$

Substituting $y = x^2$ into this equation we have

$$y = m(x - c)$$
$$x^2 = m(x - c)$$
$$x^2 - mx + mc = 0$$

Notice that the solutions of the last equation are the x-coordinates of the points of intersection of the parabola $y = x^2$ and the line l. Therefore, these solutions should be exactly $x = a$ and $x = b$.

From the Vieta's Formulas we have

$$a + b = m$$
$$ab = mc$$

Substituting $m = a + b$ into the second equation and dividing both sides by abc we have

$$ab = mc$$
$$ab = (a + b)c$$
$$ab = ac + bc$$
$$\frac{ab}{abc} = \frac{ac}{abc} + \frac{bc}{abc}$$
$$\frac{1}{c} = \frac{1}{b} + \frac{1}{a}$$

which is what needed to be proven.

Problem 3

Given a positive integer n and the polynomials

$$f_i(x) = x^2 + a_i x + a_{i+1} + 1$$

with real roots and coefficients, where $i = 1, 2, \ldots, 2n$ and $a_{2n+1} = a_1$. Prove that

$$a_1^2 + a_2^2 + \ldots + a_{2n}^2 \geq n$$

Solution

Let x_i and y_i be the roots of the polynomial $f_i(x)$. Then from the Vieta's Formulas we have

$$x_i + y_i = -a_i$$
$$x_i y_i = a_{i+1} + 1$$

From here

$$a_i^2 = (x_i + y_i)^2$$
$$a_{i+1}^2 = (x_i y_i - 1)^2$$

Now let us consider the expressions for $a_i^2 + a_{i+1}^2$:

$$
\begin{aligned}
a_i^2 + a_{i+1}^2 &= (x_i + y_i)^2 + (x_i y_i - 1)^2 \\
&= x_i^2 + 2x_i y_i + y_i^2 + x_i^2 y_i^2 - 2x_i y_i + 1 \\
&= x_i^2 + y_i^2 + x_i^2 y_i^2 + 1 \\
&= \left(x_i^2 + 1\right)\left(y_i^2 + 1\right)
\end{aligned}
$$

Notice that since x_i and y_i are real numbers then

$$\left(x_i^2 + 1\right)\left(y_i^2 + 1\right) \geq (0 + 1)(0 + 1) = 1$$

and, therefore

$$a_i^2 + a_{i+1}^2 \geq 1$$

for all $i = 1, 2, \ldots, 2n$.

Adding these $2n$ inequalities we have

$$\sum_{i=1}^{2n} \left(a_i^2 + a_{i+1}^2\right) \geq 2n$$

$$\sum_{i=1}^{2n} a_i^2 + \sum_{i=2}^{2n+1} a_i^2 \geq 2n$$

$$\sum_{i=1}^{2n} a_i^2 + \sum_{i=1}^{2n} a_i^2 \geq 2n$$

$$2\sum_{i=1}^{2n} a_i^2 \geq 2n$$

$$\sum_{i=1}^{2n} a_i^2 \geq n$$

which is what needed to be proven.

CHAPTER 14

Vieta's Formulas for Cubic Polynomial

Vieta's Formulas for Cubic Polynomial are a set of relations that establish connections between the coefficients of a cubic polynomial and its roots. These formulas are particularly useful when we need to deal with sums, products and sums of mixed products of roots of cubic polynomials.

Vieta's Formulas for Cubic Polynomial

If x_1, x_2 and x_3 are the roots of the polynomial

$$f(x) = x^3 + ax^2 + bx + c$$

then the following equalities hold

$$x_1 + x_2 + x_3 = -a$$
$$x_1 \cdot x_2 + x_2 \cdot x_3 + x_3 \cdot x_1 = b$$
$$x_1 \cdot x_2 \cdot x_3 = -c$$

Let us consider several problems.

Problem 1

Given the real numbers a, b, c, such that

$$a + b + c > 0$$
$$ab + bc + ac > 0$$
$$abc > 0$$

Prove that

$$\frac{a^2}{b} + \frac{b^2}{c} + \frac{c^2}{a} > a + b + c$$

Solution

Let us consider the polynomial $P(x)$ defined as

$$P(x) = (x - a)(x - b)(x - c)$$

which has its roots at $x = a$, $x = b$, $x = c$.

From the Vieta's Formulas we have

$$P(x) = x^3 - (a + b + c)x^2 + (ab + bc + ac)x - abc$$

Notice that if $x \leq 0$, then $x^3 \leq 0$ and

$$P(x) = x^3 - (a + b + c)x^2 + (ab + bc + ac)x - abc < 0$$

This implies that for $x \leq 0$ the value of the polynomial is negative and cannot be zero. Therefore, the roots of the polynomial are positive numbers and $a > 0$, $b > 0$, $c > 0$.

Since a, b, c are positive, then we can apply Titu's Lemma as follows:[1]

$$\frac{a^2}{b} + \frac{b^2}{c} + \frac{c^2}{a} \geq \frac{(a + b + c)^2}{a + b + c} = a + b + c$$

which is what needed to be proven.

---

[1] This inequality is discussed in detail in Chapter 26 "Titu's Lemma".

Problem 2

Real numbers x, y and z satisfy the equations

$$x + y + z = a$$

$$\frac{1}{x} + \frac{1}{y} + \frac{1}{z} = \frac{1}{a}$$

Prove that at least one of the number is equal to a.

Solution

Let us put $xy + yz + zx = b$ and rewrite the second equation as follows:

$$\frac{1}{x} + \frac{1}{y} + \frac{1}{z} = \frac{1}{a}$$

$$\frac{xy + yz + zx}{xyz} = \frac{1}{a}$$

$$a(xy + yz + zx) = xyz$$

$$ab = xyz$$

Now let us consider the polynomial $P(t)$ defined as

$$P(t) = (t - x)(t - y)(t - z)$$

which has its roots at $t = x, t = y, t = z$.

From the Vieta's Formulas we have

$$P(t) = t^3 - (x + y + z)t^2 + (xy + yz + zx)t - xyz$$

Substituting $xy + yz + zx = b$ and $xyz = ab$ we can write the polynomial $P(x)$ as follows:

$$P(t) = t^3 - at^2 + bt - ab$$

It is not hard to see that the polynomial can be factored as

$$P(t) = t^3 - at^2 + bt - ab$$
$$= t^2(t - a) + b(t - a)$$
$$= (t - a)\left(t^2 + b\right)$$

This implies that $t = a$ is a real root of the polynomial and, therefore, at least one of the numbers x, y or z is equal to a, as desired.

Problem 3

Integer numbers satisfy the condition $a + b + c = 0$. Prove that

$$2a^4 + 2b^4 + 2c^4$$

is a perfect square.

Solution

Even though this problem involves integer numbers, its solution is very algebraic.

Let us start by squaring both sides of the given equality:

$$(a + b + c)^2 = (0)^2$$
$$a^2 + b^2 + c^2 + 2(ab + bc + ac) = 0$$
$$a^2 + b^2 + c^2 = -2(ab + bc + ac)$$

Let us now consider the polynomial $P(x)$ defined as

$$P(x) = (x - a)(x - b)(x - c)$$

which has its roots at $x = a$, $x = b$, $x = c$.

From the Vieta's Formulas we have

$$P(x) = x^3 + (ab + bc + ac)x - abc$$

Since $P(a) = 0$, then we have

$$a^3 + (ab + bc + ac)a - abc = 0$$

or equivalently

$$a^3 = abc - a(ab + bc + ac)$$

Let us now multiply this equality by $2a$:

$$2a^4 = 2a^2bc - 2a^2(ab + bc + ac)$$

Similarly, we can show that

$$2b^4 = 2b^2ac - 2b^2(ab + bc + ac)$$
$$2c^4 = 2c^2ab - 2c^2(ab + bc + ac)$$

Adding the last three equalities we have

$$2a^4 + 2b^4 + 2c^4 = 2abc(a + b + c) - 2\left(a^2 + b^2 + c^2\right)(ab + bc + ac)$$

Notice that since $a + b + c = 0$ and $a^2 + b^2 + c^2 = -2(ab + bc + ac)$, then we have

$$2a^4 + 2b^4 + 2c^4 = (2ab + 2bc + 2ac)^2$$

which is what needed to be proven.[2]

---

[2]We solved this problem in a different way in Chapter 9 "Advanced Fundamental Identities".

CHAPTER 15

Roots of Polynomials

Polynomials are a fundamental concept in algebra and are used to represent a wide range of mathematical relationships. Numerous math olympiad problems revolve around the **roots of polynomials**.

Polynomial

Polynomial is a function of the form

$$f(x) = a_n x^n + a_{n-1} x^{n-1} + \ldots + a_1 x + a_0$$

where n is a nonnegative integer and a_i are real numbers.

Numbers a_i are called the **coefficients** of the polynomial $f(x)$. If all the coefficients a_i are zero, then the polynomial $f(x)$ is called **zero polynomial**. If $a_n \neq 0$, then n is called the **degree** of the polynomial $f(x)$ and a_n is called its **leading coefficient**.

By Roman Kvasov, Ph.D. Copyright © 2023 42 Points.

Root of Polynomial

Number $x = c$ is called the **root** of the polynomial $f(x)$ if $f(c) = 0$.

Fundamental Theorem of Algebra

Any polynomial of degree n has exactly n roots (some of which might be complex).

Complex Conjugate Theorem

If the complex number $a + bi$ is a root of the polynomial $f(x)$, then the complex number $a - bi$ is also a root of the polynomial $f(x)$.

Corollary for Odd-Degree Polynomials

If the polynomial $f(x)$ is of odd degree, then $f(x)$ has at least one real root.

Let us consider several problems that involve the roots of the polynomials.

Problem 1

Given the polynomials $P(x)$ and $Q(x)$ of degree 10 with their leading coefficients equal to 1. It is known that the equation $P(x) = Q(x)$ does not have real solutions. Prove that the equation

$$P\left(x^3 + 1\right) = Q\left(x^3 - 1\right)$$

has at least one real solution.

Solution

Let us put

$$P(x) = x^{10} + a_9 x^9 + \ldots + a_1 x + a_0$$
$$Q(x) = x^{10} + b_9 x^9 + \ldots + b_1 x + b_0$$

We will proceed by doing the following casework:

- If the polynomial $P(x) - Q(x)$ has odd degree, then by the Corollary for Odd-Degree Polynomials it has a real root and the equation $P(x) = Q(x)$ will have a real solution. We obtained a contradiction.

- If the polynomial $P(x) - Q(x)$ has even degree, then we have

$$P(x) - Q(x) = (a_9 - b_9)\, x^9 + \ldots + (a_1 - b_1)\, x + (a_0 - b_0)$$

and the coefficient $(a_9 - b_9)$ should be zero.

Let us prove that the polynomial $P\left(x^3 + 1\right) - Q\left(x^3 - 1\right)$ is of degree 27. Indeed, since the largest present power of x is 27, then we only need to prove that x^{27} has a nonzero coefficient.

Let us first consider $P\left(x^3 + 1\right)$:

$$P\left(x^3 + 1\right) = \left(x^3 + 1\right)^{10} + a_9 \left(x^3 + 1\right)^9 + \ldots + a_1 \left(x^3 + 1\right) + a_0$$

and from the Binomial Theorem the coefficient of x^{27} is equal to $10 + a_9$.

Let us now consider $Q\left(x^3 - 1\right)$:

$$Q\left(x^3 - 1\right) = \left(x^3 - 1\right)^{10} + b_9 \left(x^3 - 1\right)^9 + \ldots + b_1 \left(x^3 - 1\right) + b_0$$

and from the Binomial Theorem the coefficient of x^{27} is equal to $b_9 - 10$.

From here the coefficient of x^{27} of the polynomial $P\left(x^3 + 1\right) - Q\left(x^3 - 1\right)$ is

$$(10 + a_9) - (b_9 - 10) = 20 + a_9 - b_9 = 20$$

Therefore, the polynomial $P\left(x^3 + 1\right) - Q\left(x^3 - 1\right)$ is of degree 27, which implies that it has at least one real root. Consequently, we conclude that the equation

$$P\left(x^3 + 1\right) = Q\left(x^3 - 1\right)$$

has at least one real solution.

Problem 2

Given the polynomial $P(x)$ of degree 23. Prove that the equation $P\left(P\left(P(x)\right)\right) = 0$ has at least as many real solutions as the equation $P(x) = 0$.

Solution

Let us start by proving the following lemma.

Lemma

Given the polynomial $P(x)$ of odd degree. The polynomial $P(P(x))$ has at least as many real roots as the polynomial $P(x)$.

Proof

Let $x_1, x_2, \ldots, x_n$ represent all the distinct real roots of the polynomial $P(x)$. It will be enough to prove that $P(P(x))$ has at least n distinct real roots.

Let us consider the equations of the form

$$P(x) = x_i$$

for $i = 1, 2, \ldots, n$. Since $P(x)$ is of odd degree, then each of these equations has at least one real solution. Let a_i be the solution of the equation $P(x) = x_i$, i.e.

$$P(a_i) = x_i$$

for $i = 1, 2, \ldots, n$. Let us show that all a_i are distinct. Indeed, if $i \neq j$, then

$$P(a_i) = x_i \neq x_j = P(a_j)$$

Notice that since for $i = 1, 2, \ldots, n$ we have

$$P(P(a_i)) = P(x_i) = 0$$

then $P(P(x))$ has at least n distinct real roots and the lemma is proven. ∎

Let the polynomial $P(x)$ have n distinct real roots. Applying the lemma we have that $P(P(x))$ has at least n distinct real roots.

Applying the lemma again we have that $P(P(P(x)))$ has at least n distinct real roots, which is what needed to be proven.

Problem 3

Let n be a positive integer greater than 1 and $P(x)$ be a polynomial with real coefficients. Given the real numbers $a_1, a_2, \ldots, a_n, a_{n+1}$, and $b_1, b_2, \ldots, b_n, b_{n+1}$, such that $a_i \neq 0$ for $i = 1, 2, \ldots, n, n+1$. It is known that for all real values of x

$$\sum_{i=1}^{n} P(a_i x + b_i) = P(a_{n+1} x + b_{n+1})$$

Determine if the equation $P(x) = 0$ has real solutions.

Solution

We will solve this problem by doing the following casework:

- If the polynomial $P(x)$ is a constant polynomial, i.e. $P(x) = c$, then the equation becomes

$$cn = c$$
$$c(n - 1) = 0$$

Since $n > 1$, then $c = 0$ and the equation $P(x) = 0$ has real solutions.

- If the polynomial $P(x)$ is not a constant polynomial, then let us assume that the equation $P(x) = 0$ does not have real solutions. This implies that the polynomial $P(x)$ is of even degree. This in turn implies that the polynomial $P(x)$ has an absolute minimum.

Let x_0 be a point where the absolute minimum of the polynomial $P(x)$ is reached and let $P(x_0) = m$. Then

$$P(x) \geq P(x_0) = m$$

for all $x \in \mathbb{R}$.

Let us consider the value

$$y = \frac{x_0 - b_{n+1}}{a_{n+1}}$$

Notice that

$$P(a_{n+1}y + b_{n+1}) = P\left(a_{n+1} \cdot \frac{x_0 - b_{n+1}}{a_{n+1}} + b_{n+1}\right)$$
$$= P(x_0 - b_{n+1} + b_{n+1})$$
$$= P(x_0)$$
$$= m$$

Taking into account the original equation we have

$$m = P(a_{n+1}y + b_{n+1}) = \sum_{i=1}^{n} P(a_iy + b_i) \geq \sum_{i=1}^{n} m = mn > m$$

and we obtained a contradiction.

We conclude that the equation $P(x) = 0$ always has real solutions.

CHAPTER 16

Equations in Several Variables

Many math olympiad problems involve **equations in several variables**, where we either have to find the values of the variables, construct the solutions or show that the solutions do not exist. It is customary to employ casework and inequalities in these types of problems.

Let us consider several problems.

Problem 1

Is it true that for all integers $n \geq 2$, there always exist real numbers $x_1, x_2, \ldots, x_n$, such that

$$x_1 \cdot x_2 \cdot \ldots \cdot x_n = \frac{1}{1 - x_1} \cdot \frac{1}{1 - x_2} \cdot \ldots \cdot \frac{1}{1 - x_n}$$

Solution

Answer: the statement is true.

Notice that the equation

$$x = \frac{1}{1-x}$$

is equivalent to

$$x^2 - x + 1 = 0$$

and does not have real solutions.

However, the equation

$$x = -\frac{1}{1-x}$$

is equivalent to

$$x^2 - x - 1 = 0$$

and does have real solutions:

$$x = \frac{1 \pm \sqrt{5}}{2}$$

We will proceed by doing the following casework:

- If $n = 3$, then the triple $\left(2, 2, \frac{1}{2}\right)$ satisfies the conditions of the problem.

- If n is even, then we can organize the numbers $x_1, x_2, \ldots, x_n$ in pairs (x, y) of solutions of the last equation, and, therefore

$$x \cdot y = \left(-\frac{1}{1-x}\right) \cdot \left(-\frac{1}{1-y}\right) = \frac{1}{1-x} \cdot \frac{1}{1-y}$$

- If n is odd and $n \geq 5$, then it can be written in the form $n = 2k + 3$ for some $k \in \mathbb{N}$, and we can organize the numbers x_1, x_2, x_3 apart, and the rest of the numbers in pairs (x, y) of solutions like in the previous case.

We conclude that for any $n \geq 2$ there always exist real numbers $x_1, x_2, \ldots, x_n$ that satisfy the conditions of the problem.

Problem 2

For all positive integers n, find all positive real numbers $x_1, x_2, \ldots, x_n$, such that

$$(1 - x_1)^2 + (x_1 - x_2)^2 + \ldots + (x_{n-1} - x_n)^2 + x_n^2 = \frac{1}{n+1}$$

Solution

Start by noticing that the sum of the numbers under the squares is equal to

$$(1 - x_1) + (x_1 - x_2) + \ldots + (x_{n-1} - x_n) + x_n = 1$$

Let A represent the left-hand side of the equation. Let us rewrite A as follows

$$A = \frac{(1-x_1)^2}{1} + \frac{(x_1-x_2)^2}{1} + \ldots + \frac{(x_{n-1}-x_n)^2}{1} + \frac{x_n^2}{1}$$

and apply the Titu's Lemma:[1]

$$A \geq \frac{\left((1-x_1) + (x_1-x_2) + \ldots + (x_{n-1}-x_n) + x_n\right)^2}{1+1+\ldots+1+1} = \frac{1}{n+1}$$

Since the equality in Titu's Lemma holds when the variables are proportional, then we have

$$1 - x_1 = x_1 - x_2 = \ldots = x_{n-1} - x_n = x_n$$

Since the sum of these numbers is 1, then each of the numbers above is equal to $\frac{1}{n+1}$.

From here we have the following values of the variables x_i:

$$x_1 = 1 - \frac{1}{n+1}$$

$$x_2 = 1 - \frac{2}{n+1}$$

$$\ldots \qquad \ldots$$

$$x_{n-1} = 1 - \frac{n-1}{n+1}$$

$$x_n = 1 - \frac{n}{n+1}$$

These values of variables represent the only solution of the original equation.

Problem 3

Are there numbers $x_1, x_2, \ldots, x_{99}$ each equal to either $\sqrt{2} - 1$ or $\sqrt{2} + 1$, and which satisfy the equality

$$x_1x_2 + x_2x_3 + \ldots + x_{98}x_{99} + x_{99}x_1 = 199$$

Solution

Let us assume that such numbers exist.

---

[1] This inequality is discussed in detail in Chapter 26 "Titu's Lemma".

Start by noticing that the products $x_i x_{i+1}$ can only take the following values:

$$\left(\sqrt{2}-1\right)^2 = 3 - 2\sqrt{2}$$

$$\left(\sqrt{2}+1\right)^2 = 3 + 2\sqrt{2}$$

$$\left(\sqrt{2}+1\right)\left(\sqrt{2}-1\right) = 1$$

Let the number of the products that equal to $3-2\sqrt{2}$ be A, the number of the products that equal to $3+2\sqrt{2}$ be B and the number of the products that equal to 1 be C. Then we have

$$A + B + C = 99$$

Also

$$\left(3 - 2\sqrt{2}\right)A + \left(3 + 2\sqrt{2}\right)B + C = 199$$

$$3A + 3B + 2\sqrt{2}\left(B - A\right) + C = 199$$

$$2\sqrt{2}\left(B - A\right) = 199 - 3A - 3B - C$$

We will proceed by doing the following casework:

- If $B - A \neq 0$, then we have

$$\sqrt{2} = \frac{199 - 3A - 3B - C}{2\left(B - A\right)}$$

  Since the left-hand side represents an irrational number, while the right-hand side represents a rational number, then we obtained a contradiction.

- If $B - A = 0$, then we have the following system of equations:

$$\begin{cases} B - A = 0 \\ 199 - 3A - 3B - C = 0 \\ A + B + C = 99 \end{cases}$$

which implies that $A = 25$, $B = 25$ and $C = 49$.

However, the existence of the values of A, B and C does not necessarily imply that the original equation has the desired solutions.

Indeed, let us now consider the expression

$$x_1 x_2 + x_2 x_3 + \ldots + x_{98} x_{99} + x_{99} x_1$$

Let us also consider the product of all its terms:

$$x_1 x_2 \cdot x_2 x_3 \cdot \ldots \cdot x_{98} x_{99} \cdot x_{99} x_1 = \left(x_1 \cdot x_2 \cdot \ldots \cdot x_{99} \right)^2$$

Since 99 is odd, then we cannot pair $\sqrt{2} - 1$ with $\sqrt{2} + 1$ and this product is an irrational number.

On another hand it should be equal to the product of 25 terms of the form $3 - 2\sqrt{2}$, 25 terms of the form $3 + 2\sqrt{2}$ and 49 terms that equal 1, which is

$$\left(3 - 2\sqrt{2} \right)^{25} \cdot \left(3 + 2\sqrt{2} \right)^{25} \cdot 1^{49} = 1$$

and, therefore, we obtained a contradiction.

We conclude that such numbers do not exist.

CHAPTER 17

Systems of Two Equations

Many math olympiad problems involve **systems of two equations** in real numbers, where we either have to find the values of the variables, construct the solutions or show that the solutions do not exist. It is customary to employ casework and inequalities in these types of problems.

Let us consider several problems.

Problem 1

Given a positive integer $n \geq 2$ and real numbers x_1, x_2, ..., x_n that satisfy the equations:

$$\begin{cases} \dfrac{x_1}{1-x_1} + \dfrac{x_2}{1-x_2} + \ldots + \dfrac{x_n}{1-x_n} = n \\ x_1 + x_2 + \ldots + x_n = 3n \end{cases}$$

Prove that

$$\frac{x_1^2}{x_1 - 1} + \frac{x_2^2}{x_2 - 1} + \ldots + \frac{x_n^2}{x_n - 1} = \frac{1}{1 - x_1} + \frac{1}{1 - x_2} + \ldots + \frac{1}{1 - x_n}$$

Solution

Notice that from the first equation we have

$$\frac{2x_1}{1-x_1} + \frac{2x_2}{1-x_2} + \ldots + \frac{2x_n}{1-x_n} = 2n$$

Let us now introduce the following variables:

$$A = \frac{1}{1-x_1} + \frac{1}{1-x_2} + \ldots + \frac{1}{1-x_n}$$

$$B = \frac{x_1^2}{1-x_1} + \frac{x_2^2}{1-x_2} + \ldots + \frac{x_n^2}{1-x_n}$$

It will be enough to prove that $A + B = 0$.

Let us consider the expression

$$A - 2n + B = \sum_{i=1}^{n} \frac{1}{1-x_i} - \sum_{i=1}^{n} \frac{2x_i}{1-x_i} + \sum_{i=1}^{n} \frac{x_i^2}{1-x_i}$$

$$= \sum_{i=1}^{n} \frac{1 - 2x_i + x_i^2}{1-x_i}$$

$$= \sum_{i=1}^{n} \frac{(1-x_i)^2}{1-x_i}$$

$$= \sum_{i=1}^{n} (1-x_i)$$

$$= n - \sum_{i=1}^{n} x_i$$

$$= n - 3n$$

$$= -2n$$

From here we have

$$A - 2n + B = -2n$$

which implies that $A + B = 0$, as desired.

Problem 2

Find all positive integers n, such that there exist positive real numbers $x_1, x_2, \ldots, x_n$ that satisfy the equations:

$$\begin{cases} x_1 + x_2 + \ldots + x_n = 9 \\ \dfrac{1}{x_1} + \dfrac{1}{x_2} + \ldots + \dfrac{1}{x_n} = 1 \end{cases}$$

Solution

Answer: $n = 2, 3$.

By multiplying the equations and applying the AM-GM Inequality we have

$$9 = (x_1 + x_2 + \ldots + x_n)\left(\frac{1}{x_1} + \frac{1}{x_2} + \ldots + \frac{1}{x_n}\right)$$

$$\geq n \sqrt[n]{x_1 x_2 \ldots x_n} \cdot \frac{n}{\sqrt[n]{x_1 x_2 \ldots x_n}}$$

$$= n^2$$

From here we have that $n \leq 3$.

We will proceed by doing the following casework:

- If $n = 1$, then $x_1 = 9$, while $\frac{1}{x_1} = 1$, which leads to a contradiction.

- If $n = 2$, then $x_1 + x_2 = 9$, while $\frac{1}{x_1} + \frac{1}{x_2} = 1$. The last equality is equivalent to the following:

$$\frac{1}{x_1} + \frac{1}{x_2} = 1$$

$$\frac{x_1 + x_2}{x_1 x_2} = 1$$

$$\frac{9}{x_1 x_2} = 1$$

$$9 = x_1 x_2$$

Therefore, by the Vieta's Formulas[1] the numbers x_1 and x_2 are the roots of the quadratic equation $t^2 - 9t + 9 = 0$, which has the solutions $t = \frac{9 \pm 3\sqrt{5}}{2}$.

- If $n = 3$, then it is not hard to see that $x_1 = 1$, $x_2 = 1$, $x_3 = 1$ satisfy the conditions of the problem.

We conclude that only for $n = 2$ and $n = 3$ there exist positive real numbers that satisfy the conditions of the problem.

Problem 3

For all positive integer numbers n and k, determine the tuples of nonnegative real numbers $(x_1, x_2, \ldots, x_n)$ that satisfy the equations:

$$\begin{cases} x_1^k + x_2^k + \ldots + x_n^k = 1 \\ (x_1 + 1)(x_2 + 1)\ldots(x_n + 1) = 2 \end{cases}$$

---

[1] These formulas are discussed in detail in Chapter 13 "Vieta's Formulas for Quadratic Polynomial".

Solution

Answer: the solutions are all tuples $(x_1, x_2, \ldots, x_n)$, where all x_i are equal to 0, except for exactly one x_j that is equal to 1.

Start by noticing that since $x_i \geq 0$, then for all i:

$$1 = x_1^k + x_2^k + \ldots + x_n^k \geq x_i^k$$

and, therefore, $x_i \leq 1$.

From the second equation we have

$$2 = (x_1 + 1)(x_2 + 1) \ldots (x_n + 1)$$

$$\geq \sum_{1 \leq i < j \leq n} x_i x_j + (x_1 + x_2 + \ldots + x_n) + 1$$

$$\geq \sum_{1 \leq i < j \leq n} x_i x_j + \left(x_1^k + x_2^k + \ldots + x_n^k \right) + 1$$

$$\geq \sum_{1 \leq i < j \leq n} x_i x_j + 2$$

From here

$$\sum_{1 \leq i < j \leq n} x_i x_j \leq 0$$

We will proceed by doing the following casework:

- If all numbers x_i equal zero, then the first equation leads to a contradiction.

- If all numbers x_i equal zero, except for only one number x_j, then from the first equation we have $x_j = 1$ and $x_i = 0$ for all $i \neq j$.

- If all numbers x_i equal zero, except for two or more numbers, let us say x_j and x_k, then

$$\sum_{1 \leq i < j \leq n} x_i x_j \geq x_j x_k > 0$$

which leads to contradiction.

CHAPTER 18

Systems of Three Equations

Many math olympiad problems involve **systems of three equations** in real numbers. It is customary to combine the equations either by addition, subtraction or multiplication, employ casework and inequalities in these types of problems.

Let us consider several problems.

Problem 1

Solve the system of equations in positive real numbers

$$\begin{cases} 2y^2 - z = x^3 \\ 2z^2 - x = y^3 \\ 2x^2 - y = z^3 \end{cases}$$

Solution

Let us rewrite the system as

$$\begin{cases} 2y^2 = x^3 + z \\ 2z^2 = y^3 + x \\ 2x^2 = z^3 + y \end{cases}$$

Let us now apply the AM-GM Inequality to the right-hand side of each equation:

$$x^3 + z \geq 2\sqrt{x^3 z}$$
$$y^3 + x \geq 2\sqrt{y^3 x}$$
$$z^3 + y \geq 2\sqrt{z^3 x}$$

Multiplying these inequalities we have

$$\left(x^3 + z\right)\left(y^3 + x\right)\left(z^3 + y\right) \geq 8\sqrt{x^4 y^4 z^4} = 8x^2 y^2 z^2$$

and, therefore

$$8x^2 y^2 z^2 = 2y^2 \cdot 2z^2 \cdot 2x^2 = \left(x^3 + z\right)\left(y^3 + x\right)\left(z^3 + y\right) \geq 8x^2 y^2 z^2$$

Since in the AM-GM Inequality the equality holds when the variables are equal, then we have a new system of equations:

$$\begin{cases} x^3 = z \\ y^3 = x \\ z^3 = y \end{cases}$$

From here

$$z = x^3 = (y^3)^3 = y^9 = (z^3)^9 = z^{27}$$

and $z = 0$ or $z = 1$.

If $z = 0$, then $x = y = 0$.

If $z = 1$, then $x = y = 1$.

Problem 2

Solve the system of equations in real numbers

$$\begin{cases} \dfrac{1}{x} = y + z \\ \dfrac{1}{y} = z + x \\ \dfrac{1}{z} = x + y \end{cases}$$

Solution

Start by noticing that x, y and z are not equal to zero.

We will proceed by doing the following casework:

- If some two numbers x, y or z are equal, then without loss of generality we can assume that $y = z$. The first and the second equations will now form a new system of equations:

$$\begin{cases} \dfrac{1}{x} = 2y \\ \dfrac{1}{y} = y + x \end{cases}$$

  Multiplying the first equation by x and the second equation by $2y$ we have

$$\begin{cases} 1 = 2xy \\ 2 = 2y^2 + 2xy \end{cases}$$

  Subtracting these equations we have

$$1 = 2y^2$$

which implies that $y = \pm\frac{\sqrt{2}}{2}$.

If $y = \frac{\sqrt{2}}{2}$, then $x = z = \frac{\sqrt{2}}{2}$.

If $y = -\frac{\sqrt{2}}{2}$, then $x = z = -\frac{\sqrt{2}}{2}$.

- If no two numbers x, y or z are equal, then let us subtract the first two equations

$$\frac{1}{x} - \frac{1}{y} = y - x$$

$$\frac{y - x}{xy} = y - x$$

$$\frac{1}{xy} = 1$$

$$\frac{1}{x} = y$$

and, therefore, from the first equation $z = 0$, and we obtained a contradiction.

We conclude that the only solutions are the tuples

$$\left(\frac{\sqrt{2}}{2}, \frac{\sqrt{2}}{2}, \frac{\sqrt{2}}{2} \right) \text{ or } \left(\frac{\sqrt{2}}{2}, \frac{\sqrt{2}}{2}, \frac{\sqrt{2}}{2} \right)$$

Problem 3

Positive real numbers x, y and z satisfy

$$\begin{cases} x + y + z = a \\ x^2 + y^2 + z^2 = a^2 \\ x^3 + y^3 + z^3 = a^3 \end{cases}$$

Prove that one of the numbers x, y or z is equal to a.

Solution

Let us consider the identity

$$(x + y + z)^2 = x^2 + y^2 + z^2 + 2(xy + yz + zx)$$
$$(a)^2 = a^2 + 2(xy + yz + zx)$$
$$0 = 2(xy + yz + zx)$$
$$0 = xy + yz + zx$$

Let us now consider the identity

$$x^3 + y^3 + z^3 - 3xyz = (x + y + z)\left(x^2 + y^2 + z^2 - xy - yz - zx\right)$$
$$a^3 - 3xyz = (a)\left(a^2\right)$$
$$a^3 - 3xyz = a^3$$
$$-3xyz = 0$$
$$xyz = 0$$

Now let us consider the polynomial $P(t)$ defined as

$$P(t) = (t - x)(t - y)(t - z)$$

which has its roots at $t = x, t = y, t = z$.

From the Vieta's Formulas:[1]

$$P(t) = t^3 - (x + y + z)t^2 + (xy + yz + zx)t - xyz$$

Substituting $x + y + z = a$, $xy + yz + zx = 0$ and $xyz = 0$ we can write the polynomial $P(t)$ as follows:

$$P(t) = t^3 - at^2 = t^2(t - a)$$

The roots of this polynomial are $t = 0$ and $t = a$, which implies that one of the numbers x, y or z is equal to a, as desired.

---

[1] These formulas are discussed in detail in Chapter 14 "Vieta's Formulas for Cubic Polynomial".

CHAPTER 19

AM-GM Inequality Applied to Denominators

It is important to highlight that **AM-GM Inequality** is usually applied in the following equivalent form:

$$a_1 + a_2 + \ldots + a_n \geq n \sqrt[n]{a_1 \cdot a_2 \cdot \ldots \cdot a_n}$$

In this chapter we will focus on applying the AM-GM Inequality to the denominators of the fractions.

Let us consider several problems.

Problem 1

Prove that for $a, b, c > 0$ it holds that

$$\frac{1}{a} + \frac{1}{b} + \frac{1}{c} \geq \frac{a+b}{a^2+b^2} + \frac{b+c}{b^2+c^2} + \frac{c+a}{c^2+a^2}$$

Solution

Let us apply the AM-GM Inequality to each of the following denominators:

$$a^2 + b^2 \geq 2\sqrt{a^2 b^2} = 2ab$$

$$b^2 + c^2 \geq 2\sqrt{b^2 c^2} = 2bc$$

$$c^2 + a^2 \geq 2\sqrt{c^2 a^2} = 2ca$$

From here

$$\frac{a+b}{a^2+b^2} + \frac{b+c}{b^2+c^2} + \frac{c+a}{c^2+a^2} \leq \frac{a+b}{2ab} + \frac{b+c}{2bc} + \frac{c+a}{2ca}$$

$$= \frac{a}{2ab} + \frac{b}{2ab} + \frac{b}{2bc} + \frac{c}{2bc} + \frac{c}{2ca} + \frac{a}{2ca}$$

$$= \frac{1}{2b} + \frac{1}{2a} + \frac{1}{2c} + \frac{1}{2b} + \frac{1}{2a} + \frac{1}{2c}$$

$$= \frac{1}{a} + \frac{1}{b} + \frac{1}{c}$$

which is what needed to be proven.

Problem 2

Prove the inequality for positive values of a, b, c, such that $a + b + c = 1$:

$$\frac{ab}{b^3+b+2} + \frac{bc}{c^3+c+2} + \frac{ca}{a^3+a+2} \leq \frac{1}{4}$$

Solution

Let us apply the AM-GM Inequality to each of the following denominators:

$$a^3 + a + 2 = a^3 + a + 1 + 1 \geq 4\sqrt[4]{a^4} = 4a$$

$$b^3 + b + 2 = b^3 + b + 1 + 1 \geq 4\sqrt[4]{b^4} = 4b$$

$$c^3 + c + 2 = c^3 + c + 1 + 1 \geq 4\sqrt[4]{c^4} = 4c$$

From here

$$\frac{ab}{b^3 + b + 2} + \frac{bc}{c^3 + c + 2} + \frac{ca}{a^3 + a + 2} \leq \frac{ab}{4b} + \frac{bc}{4c} + \frac{ca}{4a}$$

$$= \frac{a}{4} + \frac{b}{4} + \frac{c}{4}$$

$$= \frac{a + b + c}{4}$$

$$= \frac{1}{4}$$

which is what needed to be proven.

Problem 3

Given positive real numbers a, b and c. Prove that

$$\sum_{\text{cyc}} \frac{2 + a^2}{\sqrt{1 + b^3}} \geq 6$$

Solution

Let us factor the expression under the root and apply the AM-GM Inequality:

$$\sqrt{1 + b^3} = \sqrt{(1 + b)(1 - b + b^2)}$$

$$\leq \frac{(1 + b) + (1 - b + b^2)}{2}$$

$$= \frac{2 + b^2}{2}$$

From here

$$\sum_{\text{cyc}} \frac{2 + a^2}{\sqrt{1 + b^3}} \geq \sum_{\text{cyc}} \frac{2(2 + a^2)}{2 + b^2} = 2 \cdot \sum_{\text{cyc}} \frac{2 + a^2}{2 + b^2}$$

However, by the AM-GM Inequality we have

$$2 \cdot \sum_{\text{cyc}} \frac{2+a^2}{2+b^2} \geq 2 \cdot 3 \sqrt[3]{\frac{2+a^2}{2+b^2} \cdot \frac{2+b^2}{2+c^2} \cdot \frac{2+c^2}{2+a^2}} = 6$$

as desired.

CHAPTER 20

AM-GM Inequality and Cyclic Permutations

In this chapter we will look at several math olympiad problems that involve the concept of **AM-GM Inequality**. We will focus on applying the **AM-GM Inequality** to some parts of the original inequality and then combine the obtained inequalities. We recommend reviewing the main ideas discussed in Chapter 19 "AM-GM Inequality Applied to Denominators" before working on the exercises from this chapter.

Let us consider several problems.

Problem 1

Positive numbers a, b and c are such that $a + b + c = 1$. Prove the inequality

$$a^2 + b^2 + c^2 + \frac{4}{a(a+1)} + \frac{4}{b(b+1)} + \frac{4}{c(c+1)} \geq 11$$

Solution

Notice that applying the AM-GM Inequality to $a(a + 1)$ and $\frac{4}{a(a+1)}$ we have

$$a(a + 1) + \frac{4}{a(a + 1)} \geq 4$$

Similarly, we can show that

$$b(b + 1) + \frac{4}{b(b + 1)} \geq 4$$

$$c(c + 1) + \frac{4}{c(c + 1)} \geq 4$$

by cyclically permuting the variables.

Adding these inequalities and taking into account that $a + b + c = 1$ we have

$$a(a + 1) + \frac{4}{a(a + 1)} + b(b + 1) + \frac{4}{b(b + 1)} + c(c + 1) + \frac{4}{c(c + 1)} \geq 12$$

$$a^2 + b^2 + c^2 + \frac{4}{a(a + 1)} + \frac{4}{b(b + 1)} + \frac{4}{c(c + 1)} + a + b + c \geq 12$$

$$a^2 + b^2 + c^2 + \frac{4}{a(a + 1)} + \frac{4}{b(b + 1)} + \frac{4}{c(c + 1)} \geq 11$$

which is what needed to be proven.

Problem 2

Given the positive real numbers a, b, c, such that

$$ab + bc + ac = 5$$

$$a^2 + b^2 + c^2 = 6$$

Prove that

$$\frac{a^4}{b} + \frac{b^4}{c} + \frac{c^4}{a} \geq 8$$

Solution

First, notice that we can find the value of $a + b + c$ from

$$(a + b + c)^2 = a^2 + b^2 + c^2 + 2(ab + bc + ac) = 6 + 2(5) = 16$$

Since the numbers a, b, c are positive, then $a + b + c = 4$.

Now we will apply the AM-GM Inequality to $\frac{a^4}{b}$ and b:

$$\frac{a^4}{b} + b \geq 2\sqrt{a^4} = 2a^2$$

Similarly, we can show that

$$\frac{b^4}{c} + c \geq 2\sqrt{b^4} = 2b^2$$

$$\frac{c^4}{a} + a \geq 2\sqrt{c^4} = 2c^2$$

by cyclically permuting the variables.

Adding these inequalities and taking into account that we have $a + b + c = 4$ and $a^2 + b^2 + c^2 = 6$ implies that

$$\frac{a^4}{b} + b + \frac{b^4}{c} + c + \frac{c^4}{a} + a \geq 2\left(a^2 + b^2 + c^2\right)$$

$$\frac{a^4}{b} + \frac{b^4}{c} + \frac{c^4}{a} + 4 \geq 2\,(6)$$

$$\frac{a^4}{b} + \frac{b^4}{c} + \frac{c^4}{a} \geq 8$$

which is what needed to be proven.

Problem 3

For $a, b, c > 0$ and $ab + bc + ac = 1$, prove that

$$\sqrt{a + \frac{1}{a}} + \sqrt{b + \frac{1}{b}} + \sqrt{c + \frac{1}{c}} \geq 2\left(\sqrt{a} + \sqrt{b} + \sqrt{c}\right)$$

Solution

Let us start by considering the expression $a + \frac{1}{a}$ and substitute 1 for $ab + bc + ac$:

$$a + \frac{1}{a} = a + \frac{ab + bc + ac}{a} = a + b + c + \frac{bc}{a}$$

Applying AM-GM Inequality to the terms a and $\frac{bc}{a}$ we have

$$a + \frac{bc}{a} \geq 2\sqrt{bc}$$

Therefore

$$a + \frac{1}{a} = \left(a + \frac{bc}{a}\right) + b + c \geq 2\sqrt{bc} + b + c = \left(\sqrt{b} + \sqrt{c}\right)^2$$

From here we have

$$\sqrt{a + \frac{1}{a}} \geq \sqrt{b} + \sqrt{c}$$

Similarly, we can show that

$$\sqrt{b + \frac{1}{b}} \geq \sqrt{c} + \sqrt{a}$$

$$\sqrt{c + \frac{1}{c}} \geq \sqrt{a} + \sqrt{b}$$

by cyclically permuting the variables.

Adding these inequalities we have

$$\sqrt{a + \frac{1}{a}} + \sqrt{b + \frac{1}{b}} + \sqrt{c + \frac{1}{c}} \geq 2\left(\sqrt{a} + \sqrt{b} + \sqrt{c}\right)$$

which is what needed to be proven.

CHAPTER 21

AM-GM Inequality for More Variables

In this chapter we will look at several math olympiad problems that involve the concept of **AM-GM Inequality**. We will focus on applying the **AM-GM Inequality** to the problems that involve n variables. We recommend reviewing the main ideas discussed in Chapter 19 "AM-GM Inequality Applied to Denominators" and Chapter 20 "AM-GM Inequality and Cyclic Permutations" before working on the exercises from this chapter.

Let us consider several problems.

Problem 1

Prove the following inequality for all integers $n \geq 2$:

$$\frac{1}{n} + \frac{1}{n+1} + \ldots + \frac{1}{2n-1} > n\left(2^{\frac{1}{n}} - 1\right)$$

Solution

Start by noticing that the original inequality is equivalent to

$$\frac{1}{n} + \frac{1}{n+1} + \ldots + \frac{1}{2n-1} > n\left(2^{\frac{1}{n}} - 1\right)$$

$$\frac{1}{n} + \frac{1}{n+1} + \ldots + \frac{1}{2n-1} > n\sqrt[n]{2} - n$$

$$\frac{1}{n} + \frac{1}{n+1} + \ldots + \frac{1}{2n-1} + n > n\sqrt[n]{2}$$

$$\left(\frac{1}{n} + 1\right) + \left(\frac{1}{n+1} + 1\right) + \ldots + \left(\frac{1}{2n-1} + 1\right) > n\sqrt[n]{2}$$

$$\frac{n+1}{n} + \frac{n+2}{n+1} + \ldots + \frac{2n}{2n-1} > n\sqrt[n]{2}$$

The last inequality follows directly from the AM-GM Inequality and telescoping the product under the radical:

$$\frac{n+1}{n} + \frac{n+2}{n+1} + \ldots + \frac{2n}{2n-1} \geq \sqrt[n]{\frac{2n}{n}} = n\sqrt[n]{2}$$

Since in our case all the variables are distinct, then the equality is not reached and

$$\frac{n+1}{n} + \frac{n+2}{n+1} + \ldots + \frac{2n}{2n-1} > n\sqrt[n]{2}$$

as desired.

Problem 2

Let $x_0 > x_1 > \ldots > x_n$ be real numbers. Prove that

$$x_0 + \frac{1}{x_0 - x_1} + \frac{1}{x_1 - x_2} + \ldots + \frac{1}{x_{n-1} - x_n} \geq x_n + 2n$$

Solution

Start by noticing that the original inequality is equivalent to

$$x_0 - x_n + \frac{1}{x_0 - x_1} + \frac{1}{x_1 - x_2} + \ldots + \frac{1}{x_{n-1} - x_n} \geq 2n$$

Let us now make the substitution $a_k = x_k - x_{k+1}$, where $k = 0, 1, \ldots, n - 1$.

Notice that

$$a_0 + a_1 + \ldots + a_{n-1} = (x_0 - x_1) + (x_1 - x_2) + \ldots + (x_{n-1} - x_n) = x_0 - x_n$$

Therefore, the initial inequality is equivalent to

$$a_0 + a_1 + \ldots + a_{n-1} + \frac{1}{a_0} + \frac{1}{a_1} + \ldots + \frac{1}{a_{n-1}} \geq 2n$$

which follows directly from the AM-GM Inequality:

$$a_0 + a_1 + \ldots + a_{n-1} + \frac{1}{a_0} + \frac{1}{a_1} + \ldots + \frac{1}{a_{n-1}} \geq 2n \sqrt[2n]{\frac{a_0 a_1 \ldots a_{n-1}}{a_0 a_1 \ldots a_{n-1}}} = 2n$$

Problem 3

Let $P(x)$ be a polynomial with real non-negative coefficients. Let $x_1, x_2, \ldots, x_n$ be positive real numbers, such that

$$x_1 \cdot x_2 \cdot \ldots \cdot x_n = 1$$

Prove that

$$P(x_1) + P(x_2) + \ldots + P(x_n) \geq nP(1)$$

Solution

Let the polynomial $P(x)$ be

$$P(x) = a_k x^k + \ldots + a_1 x^1 + a_0 x^0 = \sum_{i=0}^{k} a_i x^i$$

where $a_i \geq 0$. Then

$$P(1) = a_k + \ldots + a_1 + a_0 = \sum_{i=0}^{k} a_i$$

The left-hand side of the inequality is equal to

$$P(x_1) + P(x_2) + \ldots + P(x_n) = \sum_{i=0}^{k} a_i x_1^i + \sum_{i=0}^{k} a_i x_2^i + \ldots + \sum_{i=0}^{k} a_i x_n^i$$

However

$$\sum_{i=0}^{k} a_i x_1^i + \sum_{i=0}^{k} a_i x_2^i + \ldots + \sum_{i=0}^{k} a_i x_n^i = a_0 \sum_{i=1}^{n} x_i^0 + a_1 \sum_{i=1}^{n} x_i^1 + \ldots + a_k \sum_{i=1}^{n} x_i^k$$

Now, taking into account that $x_1 x_2 \ldots x_n = 1$, we will apply the AM-GM Inequality to each sum:

$$\sum_{i=1}^{n} x_i^0 = x_1^0 + x_2^0 + \ldots + x_n^0 = n$$

$$\sum_{i=1}^{n} x_i^1 = x_1^1 + x_2^1 + \ldots + x_n^1 \geq n$$

$$\ldots \qquad \qquad \ldots \qquad \qquad \ldots$$

$$\sum_{i=1}^{n} x_i^k = x_1^k + x_2^k + \ldots + x_n^k \geq n$$

Therefore

$$P(x_1) + P(x_2) + \ldots + P(x_n) = a_0 \sum_{i=1}^{n} x_i^0 + a_1 \sum_{i=1}^{n} x_i^1 + \ldots + a_k \sum_{i=1}^{n} x_i^k$$

$$\geq a_0 n + a_1 n + \ldots + a_k n$$

$$\geq n \left(a_0 + a_1 + \ldots + a_k \right)$$

$$\geq n P(1)$$

which is what needed to be proven.

CHAPTER 22

AM-GM Inequality and Other Ideas

In this chapter we will look at several math olympiad problems that involve the concept of **AM-GM Inequality**. We will focus on how the **AM-GM Inequality** is used together with other inequalities and techniques. We recommend reviewing the main ideas discussed in Chapter 19 "AM-GM Inequality Applied to Denominators" and Chapter 20 "AM-GM Inequality and Cyclic Permutations" before working on the exercises from this chapter.

Let us consider several problems.

Problem 1

Given the positive real numbers a, b, c, such that $abc = 1$. Prove the inequality

$$\frac{a^2b}{a+c} + \frac{b^2c}{b+a} + \frac{c^2a}{c+b} \geq \frac{3}{2}$$

Solution

Let us rewrite the left-hand side of the inequality as

$$\frac{a^2b^2}{ab+bc}+\frac{b^2c^2}{bc+ac}+\frac{c^2a^2}{ac+ab}$$

Now we will apply the Titu's Lemma[1]:

$$\frac{a^2b^2}{ab+bc}+\frac{b^2c^2}{bc+ac}+\frac{c^2a^2}{ac+ab}\geq\frac{(ab+bc+ac)^2}{2(ab+bc+ac)}=\frac{ab+bc+ac}{2}$$

Applying the AM-GM Inequality we have:

$$\frac{ab+bc+ac}{2}\geq\frac{3\sqrt[3]{a^2b^2c^2}}{2}=\frac{3}{2}$$

which is what needed to be proven.

Problem 2

Let $a,b,c>0$ and $abc=1$. Prove that

$$\frac{a}{(a+1)(b+1)}+\frac{b}{(b+1)(c+1)}+\frac{c}{(c+1)(a+1)}\geq\frac{3}{4}$$

Solution

Let $x=a+b+c$, $y=ab+bc+ac$ and z be the left-hand side of the inequality.

Let us start by adding the fractions that constitute z:

$$z=\frac{a}{(a+1)(b+1)}+\frac{b}{(b+1)(c+1)}+\frac{c}{(c+1)(a+1)}$$

$$=\frac{a(c+1)}{(a+1)(b+1)(c+1)}+\frac{b(a+1)}{(b+1)(c+1)(a+1)}+\frac{c(b+1)}{(c+1)(a+1)(b+1)}$$

$$=\frac{a+b+c+ab+bc+ac}{(a+1)(b+1)(c+1)}$$

$$=\frac{x+y}{(a+1)(b+1)(c+1)}$$

If we expand the denominator we will get

$$(a+1)(b+1)(c+1)=abc+ab+bc+ac+a+b+c+1=x+y+2$$

---

[1]This inequality is discussed in detail in Chapter 26 "Titu's Lemma".

Therefore

$$z = \frac{x+y}{x+y+2}$$

and it will be enough to prove that

$$\frac{x+y}{x+y+2} \geq \frac{3}{4}$$

This inequality is equivalent to

$$\frac{x+y}{x+y+2} \geq \frac{3}{4}$$

$$4(x+y) \geq 3(x+y+2)$$

$$x+y \geq 6$$

However, the last inequality follows directly from the AM-GM Inequality:

$$x+y = a+b+c+ab+bc+ac \geq 6\sqrt[6]{a^6 b^6 c^6} = 6$$

Problem 3

Prove that for positive a, b, c

$$\frac{a}{2b+c} + \frac{b}{2c+a} + \frac{c}{2a+b} \geq \frac{2}{3}$$

Solution

Let us start by introducing the new variables

$$x = 2b+c$$

$$y = 2c+a$$

$$z = 2a+b$$

Now we will express the variables a, b, c in terms of the variables x, y, z:

$$a = \frac{-2x+y+4z}{9}$$

$$b = \frac{-2y+z+4x}{9}$$

$$c = \frac{-2z+x+4y}{9}$$

Therefore, the inequality is equivalent to

$$\frac{-2x+y+4z}{9x} + \frac{-2y+z+4x}{9y} + \frac{-2z+x+4y}{9z} \geq \frac{2}{3}$$

$$\frac{-2x+y+4z}{x} + \frac{-2y+z+4x}{y} + \frac{-2z+x+4y}{z} \geq 6$$

$$\frac{-2x}{x} + \frac{y}{x} + \frac{4z}{x} + \frac{-2y}{y} + \frac{z}{y} + \frac{4x}{y} + \frac{-2z}{z} + \frac{x}{z} + \frac{4y}{z} \geq 6$$

$$\frac{y}{x} + \frac{4z}{x} + \frac{z}{y} + \frac{4x}{y} + \frac{x}{z} + \frac{4y}{z} \geq 12$$

However, the last inequality follows directly from the AM-GM Inequality:

$$\frac{y}{x} + \frac{4z}{x} + \frac{z}{y} + \frac{4x}{y} + \frac{x}{z} + \frac{4y}{z} \geq 6\sqrt[6]{4^3 \cdot \frac{y}{x} \cdot \frac{z}{x} \cdot \frac{z}{y} \cdot \frac{x}{y} \cdot \frac{x}{z} \cdot \frac{y}{z}} = 12$$

CHAPTER 23

Weighted AM-GM Inequality: Integer Weights

Weighted AM-GM Inequality

Weighted AM-GM Inequality states that for the positive real numbers $(a_1, a_2, \ldots, a_n)$ and the positive weights $(m_1, m_2, \ldots, m_n)$ the following inequality holds:

$$\frac{m_1 a_1 + m_2 a_2 + \ldots + m_n a_n}{m_1 + m_2 + \ldots + m_n} \geq \left(a_1^{m_1} \cdot a_2^{m_2} \cdot \ldots \cdot a_n^{m_n}\right)^{\frac{1}{m_1 + m_2 + \ldots + m_n}}$$

The equality holds when all the variables a_i are equal.

In this chapter we will see how Weighted AM-GM Inequality is deployed with weights being constant integers. This approach is equivalent to using the classical AM-GM Inequality with repeated terms.

Let us consider several problems.

Problem 1

Prove that for $x, y, z \geq 2$ the following inequality holds:

$$\left(y^3 + x\right)\left(z^3 + y\right)\left(x^3 + z\right) \geq 125xyz$$

Solution

Start by noticing that since $y \geq 2$, then

$$y^3 + x = y^2 \cdot y + x \geq 4y + x$$

Now let us apply the Weighted AM-GM Inequality to the numbers (y, x) with weights $(4, 1)$:

$$\frac{4y + x}{4 + 1} \geq \left(y^4 \cdot x^1\right)^{\frac{1}{4+1}}$$

which is equivalent to

$$\frac{4y + x}{5} \geq \left(y^4 x\right)^{\frac{1}{5}}$$

$$4y + x \geq 5\sqrt[5]{y^4 x}$$

From here we have that

$$y^3 + x \geq 4y + x \geq 5\sqrt[5]{y^4 x}$$

Similarly, we can show that

$$z^3 + y \geq 5\sqrt[5]{z^4 y}$$

$$x^3 + z \geq 5\sqrt[5]{x^4 z}$$

Therefore, by multiplying these inequalities we have

$$\left(y^3 + x\right)\left(z^3 + y\right)\left(x^3 + z\right) \geq 125\sqrt[5]{x^5 y^5 z^5} = 125xyz$$

which is what needed to be proven.

Problem 2

Let a, b, c be positive real numbers, such that $a + b + c = 3$. Prove that

$$\sqrt{a} + \sqrt{b} + \sqrt{c} \geq ab + bc + ca$$

Solution

Let us start by squaring and rewriting the given equality:

$$a + b + c = 3$$

$$(a + b + c)^2 = (3)^2$$

$$a^2 + b^2 + c^2 + 2(ab + bc + ac) = 9$$

$$2(ab + bc + ac) = 9 - a^2 - b^2 - c^2$$

$$ab + bc + ac = \frac{1}{2}\left(9 - a^2 - b^2 - c^2\right)$$

By substituting the expression for $ab + bc + ac$ into the original inequality we can rewrite it as follows:

$$\sqrt{a} + \sqrt{b} + \sqrt{c} \geq ab + bc + ca$$

$$\sqrt{a} + \sqrt{b} + \sqrt{c} \geq \frac{1}{2}\left(9 - a^2 - b^2 - c^2\right)$$

$$2\sqrt{a} + 2\sqrt{b} + 2\sqrt{c} \geq 9 - a^2 - b^2 - c^2$$

$$\left(a^2 + 2\sqrt{a}\right) + \left(b^2 + 2\sqrt{b}\right) + \left(c^2 + 2\sqrt{c}\right) \geq 9$$

Let us now prove the last inequality.

We will apply the Weighted AM-GM Inequality to the numbers $\left(a^2, \sqrt{a}\right)$ with weights $(1, 2)$:

$$\frac{1a^2 + 2\sqrt{a}}{1 + 2} \geq \left(a^2 \cdot \left(\sqrt{a}\right)^2\right)^{\frac{1}{1+2}}$$

which is equivalent to

$$\frac{a^2 + 2\sqrt{a}}{3} \geq \left(a^3\right)^{\frac{1}{3}}$$

$$a^2 + 2\sqrt{a} \geq 3a$$

Similarly, we have

$$b^2 + 2\sqrt{b} \geq 3b$$

$$c^2 + 2\sqrt{c} \geq 3c$$

By adding these inequalities we have

$$\left(a^2 + 2\sqrt{a}\right) + \left(b^2 + 2\sqrt{b}\right) + \left(c^2 + 2\sqrt{c}\right) \geq 3a + 3b + 3c$$

$$= 3(a + b + c)$$

$$= 9$$

which is what needed to be proven.

Problem 3

Given positive reals a, b and c, such that

$$a + b + c = \frac{1}{16}$$

Prove the inequality

$$\sqrt[4]{a} + \sqrt[4]{b} + \sqrt[4]{c} \le \frac{5}{4}$$

Solution

Let us start by proving the following inequality:

$$2x + \frac{3}{8} \ge \sqrt[4]{x}$$

Indeed, applying the Weighted AM-GM Inequality to the numbers $\left(2x, \frac{1}{8}\right)$ with weights $(1, 3)$ we have:

$$\frac{1 \cdot 2x + 3 \cdot \frac{1}{8}}{1 + 3} \ge \left((2x)^1 \cdot \left(\frac{1}{8}\right)^3\right)^{\frac{1}{1+3}}$$

which is equivalent to

$$\frac{2x + \frac{3}{8}}{4} \ge \left(\frac{x}{256}\right)^{\frac{1}{4}}$$

$$2x + \frac{3}{8} \ge \sqrt[4]{x}$$

Therefore we have the following inequalities for a, b and c:

$$2a + \frac{3}{8} \ge \sqrt[4]{a}$$

$$2b + \frac{3}{8} \ge \sqrt[4]{b}$$

$$2c + \frac{3}{8} \ge \sqrt[4]{c}$$

Adding these inequalities we have

$$2(a + b + c) + \frac{9}{8} \ge \sqrt[4]{a} + \sqrt[4]{b} + \sqrt[4]{c}$$

$$2\left(\frac{1}{16}\right) + \frac{9}{8} \ge \sqrt[4]{a} + \sqrt[4]{b} + \sqrt[4]{c}$$

$$\frac{5}{4} \ge \sqrt[4]{a} + \sqrt[4]{b} + \sqrt[4]{c}$$

which is what needed to be proven.

CHAPTER 24

Weighted AM-GM Inequality: Variable Weights

In this chapter we will look at several math olympiad problems that involve the concept of **Weighted AM-GM Inequality**. We will focus on how **Weighted AM-GM Inequality** is used with variables as weights. We recommend reviewing the main ideas discussed in Chapter 23 "Weighted AM-GM Inequality: Integer Weights" before working on the exercises from this chapter.

Let us consider several problems.

Problem 1

Prove that for positive a and b it holds that

$$\left(\frac{a+b}{2}\right)^{a+b} \geq a^b \cdot b^a$$

Solution

Let us apply the Weighted AM-GM Inequality to the numbers (a, b) with weights (b, a):

$$\frac{ab + ba}{a + b} \geq \left(a^b \cdot b^a\right)^{\frac{1}{a+b}}$$

which is equivalent to

$$\left(\frac{2ab}{a + b}\right)^{a+b} \geq a^b \cdot b^a$$

Let us prove that

$$\frac{a + b}{2} \geq \frac{2ab}{a + b}$$

Indeed, this inequality is equivalent to an obvious inequality:

$$\frac{a + b}{2} \geq \frac{2ab}{a + b}$$

$$(a + b)^2 \geq 4ab$$

$$a^2 + 2ab + b^2 \geq 4ab$$

$$a^2 - 2ab + b^2 \geq 0$$

$$(a - b)^2 \geq 0$$

From here we have

$$\left(\frac{a + b}{2}\right)^{a+b} \geq \left(\frac{2ab}{a + b}\right)^{a+b} \geq a^b \cdot b^a$$

which is what needed to be proven.

Problem 2

Prove that if $a, b, c > 0$ and $a + b + c = 3$, then

$$a^b b^c c^a \leq 1$$

Solution

Let us apply the Weighted AM-GM Inequality to the numbers (a, b, c) with weights (b, c, a):

$$\frac{ab + bc + ca}{a + b + c} \geq \left(a^b \cdot b^c \cdot c^a\right)^{\frac{1}{a+b+c}}$$

This inequality is equivalent to

$$\frac{ab + bc + ca}{3} \geq \sqrt[3]{a^b \cdot b^c \cdot c^a}$$

$$\left(\frac{ab + bc + ca}{3}\right)^3 \geq a^b \cdot b^c \cdot c^a$$

Notice that it will be enough to prove that

$$\frac{ab + bc + ca}{3} \leq 1$$

or equivalently

$$3 \geq ab + bc + ac$$

Let us consider an obvious inequality:

$$(a - b)^2 + (b - c)^2 + (a - c)^2 \geq 0$$

After the expansion it becomes equivalent to

$$2a^2 + 2b^2 + 2c^2 - 2(ab + bc + ac) \geq 0$$

$$a^2 + b^2 + c^2 - (ab + bc + ac) \geq 0$$

$$a^2 + b^2 + c^2 + 2(ab + bc + ac) \geq 3(ab + bc + ac)$$

$$(a + b + c)^2 \geq 3(ab + bc + ac)$$

$$(3)^2 \geq 3(ab + bc + ac)$$

$$3 \geq ab + bc + ac$$

as desired.

Problem 3

Let a, b, c, d be positive real numbers, such that

$$a + b + c + d = \sqrt[4]{a} + \sqrt[4]{b} + \sqrt[4]{c} + \sqrt[4]{d}$$

Prove that

$$a^a b^b c^c d^d \geq 1$$

Solution

Let us put $s = a + b + c + d$. Dividing both sides of this equality by s we have

$$\frac{a}{s} + \frac{b}{s} + \frac{c}{s} + \frac{d}{s} = 1$$

We will now apply the Weighted AM-GM Inequality to the numbers

$$\left(a^{-\frac{3}{4}}, b^{-\frac{3}{4}}, c^{-\frac{3}{4}}, d^{-\frac{3}{4}}\right)$$

with the following weights

$$\left(\frac{a}{s}, \frac{b}{s}, \frac{c}{s}, \frac{d}{s}\right)$$

Notice that the left-hand side of the Weighted AM-GM Inequality becomes

$$\frac{\frac{a}{s} \cdot a^{-\frac{3}{4}} + \frac{b}{s} \cdot b^{-\frac{3}{4}} + \frac{c}{s} \cdot c^{-\frac{3}{4}} + \frac{d}{s} \cdot d^{-\frac{3}{4}}}{\frac{a}{s} + \frac{b}{s} + \frac{c}{s} + \frac{d}{s}} = \frac{a^{\frac{1}{4}}}{s} + \frac{b^{\frac{1}{4}}}{s} + \frac{c^{\frac{1}{4}}}{s} + \frac{d^{\frac{1}{4}}}{s}$$

$$= \frac{a^{\frac{1}{4}} + b^{\frac{1}{4}} + c^{\frac{1}{4}} + d^{\frac{1}{4}}}{s}$$

$$= \frac{\sqrt[4]{a} + \sqrt[4]{b} + \sqrt[4]{c} + \sqrt[4]{d}}{a + b + c + d}$$

$$= 1$$

Now the right-hand side of the Weighted AM-GM Inequality becomes

$$\left(\left(a^{-\frac{3}{4}}\right)^{\frac{a}{s}} \left(b^{-\frac{3}{4}}\right)^{\frac{b}{s}} \left(c^{-\frac{3}{4}}\right)^{\frac{c}{s}} \left(d^{-\frac{3}{4}}\right)^{\frac{d}{s}}\right)^{\frac{1}{\frac{a}{s} + \frac{b}{s} + \frac{c}{s} + \frac{d}{s}}} = \left(a^a b^b c^c d^d\right)^{-\frac{3}{4s}}$$

Therefore, from the Weighted AM-GM Inequality it follows that

$$1 \geq \left(a^a b^b c^c d^d\right)^{-\frac{3}{4s}}$$

which is equivalent to

$$a^a b^b c^c d^d \geq 1$$

as desired.

CHAPTER 25

Cauchy-Schwarz Inequality

Cauchy-Schwarz Inequality

Cauchy-Schwarz Inequality states that for the real numbers $(a_1, a_2, \ldots, a_n)$ and $(b_1, b_2, \ldots, b_n)$ the following inequality holds:

$$(a_1 b_1 + a_2 b_2 + \ldots + a_n b_n)^2 \leq (a_1^2 + a_2^2 + \ldots + a_n^2) \cdot (b_1^2 + b_2^2 + \ldots + b_n^2)$$

The equality holds when the variables are proportional, i.e. if there exists a constant t such that $a_i = t b_i$ for all $i = 1, 2, \ldots, n$.

Cauchy-Schwarz Inequality is usually applied to the problems that contain the summation of mixed products.

Let us consider several problems.

Problem 1

Given that

$$x^2 + y^2 + z^2 = 1$$

Find the maximum value of the expression

$$2x + 3y + 6z$$

Solution

Let us apply the Cauchy-Schwarz Inequality to the numbers $(2, 3, 6)$ and (x, y, z):

$$(2x + 3y + 6z)^2 \leq (2^2 + 3^2 + 6^2) \cdot (x^2 + y^2 + z^2) = (49)(1) = 49$$

which implies that

$$2x + 3y + 6z \leq 7$$

Now we have to verify if the value 7 can be reached. Since in the Cauchy-Schwarz Inequality the equality holds when the variables are proportional, then we have

$$x = 2t$$

$$y = 3t$$

$$z = 6t$$

We can find the value of t by substituting the expressions for x, y and z into the equation:

$$2x + 3y + 6z = 7$$

$$2(2t) + 3(3t) + 6(6t) = 7$$

$$49t = 7$$

$$t = \frac{1}{7}$$

From here we find the values of the variables for which the maximum is reached:

$$x = \frac{2}{7}, \quad y = \frac{3}{7}, \quad z = \frac{6}{7}$$

Problem 2

Let $P(x)$ be a polynomial with nonnegative coefficients. Show that for any real numbers x and y, the following inequality holds

$$(P(xy))^2 \leq P(x^2) \cdot P(y^2)$$

Solution

Since the coefficients of the polynomial are nonnegative, then we can assume them to be $b_0^2, b_1^2, \ldots, b_n^2$ for some real numbers b_i and the polynomial $P(x)$ to be written as

$$P(x) = b_n^2 x^n + \ldots + b_1^2 x + b_0^2$$

Now let us apply the Cauchy-Schwarz Inequality to the numbers $(b_n x^n, \ldots, b_1 x, b_0)$ and $(b_n y^n, \ldots, b_1 y, b_0)$:

$$(P(xy))^2 = \left(b_n^2 (xy)^n + \ldots + b_1^2 (xy) + b_0^2\right)^2$$
$$= ((b_n x^n) \cdot (b_n y^n) + \ldots + (b_1 x) \cdot (b_1 y) + (b_0) \cdot (b_0))^2$$
$$\leq \left(b_n^2 x^{2n} + \ldots + b_1^2 x^2 + b_0^2\right) \cdot \left(b_n^2 y^{2n} + \ldots + b_1^2 y^2 + b_0^2\right)$$
$$= P\left(x^2\right) \cdot P\left(y^2\right)$$

which is what needed to be proven.

Problem 3

Prove that for all integers $n \geq 2$

$$\sqrt{1} + \sqrt{2} + \ldots + \sqrt{n^2 - 1} < \frac{n^3 - n}{\sqrt{2}}$$

Solution

Let X be the quantity defined as

$$X = \sqrt{1} + \sqrt{2} + \ldots + \sqrt{n^2 - 1}$$

Let us now apply the Cauchy-Schwarz Inequality to the numbers $\left(\sqrt{1}, \sqrt{2}, \ldots, \sqrt{n^2 - 1}\right)$ and $(1, 1, \ldots, 1)$:

$$X^2 = \left(\sqrt{1} + \sqrt{2} + \ldots + \sqrt{n^2 - 1}\right)^2$$
$$= \left(\sqrt{1} \cdot 1 + \sqrt{2} \cdot 1 + \ldots + \sqrt{n^2 - 1} \cdot 1\right)^2$$
$$\leq \left(1 + 2 + \ldots + \left(n^2 - 1\right)\right)^2 \cdot (1 + 1 + \ldots + 1)^2$$

Notice that the second parenthesis has exactly $n^2 - 1$ ones added up, while the addition in the first parenthesis can be done using the Gauss Formula:

$$X^2 \leq \left(1 + 2 + \ldots + \left(n^2 - 1\right)\right)^2 \cdot \left(1 + 1 + \ldots + 1\right)^2$$

$$X^2 \leq \frac{\left(n^2 - 1\right) \cdot n^2}{2} \cdot \left(n^2 - 1\right)^2$$

$$X^2 \leq \frac{\left(\left(n^2 - 1\right) n\right)^2}{2}$$

$$X^2 \leq \frac{\left(n^3 - n\right)^2}{2}$$

The last inequality implies that

$$X \leq \frac{n^3 - n}{\sqrt{2}}$$

Since in the Cauchy-Schwarz Inequality, the equality holds when the variables are proportional, then in our problem the equality does not hold for any integer $n \geq 2$ and

$$X < \frac{n^3 - n}{\sqrt{2}}$$

as desired.

CHAPTER 26

Titu's Lemma

Titu's Lemma

Titu's Lemma states that for positive real numbers $(a_1, a_2, \ldots, a_n)$ and $(b_1, b_2, \ldots, b_n)$ the following inequality holds:

$$\frac{a_1^2}{b_1} + \frac{a_2^2}{b_2} + \ldots + \frac{a_n^2}{b_n} \geq \frac{(a_1 + a_2 + \ldots + a_n)^2}{b_1 + b_2 + \ldots + b_n}$$

The equality holds when the variables are proportional, i.e. if there exists a constant t such that $a_i = tb_i$ for all $i = 1, 2, \ldots, n$.

Titu's Lemma is usually applied to the problems that contain the summation of several fractions with numerators containing squares. If the squares do not appear in the numerators they can be obtained by squaring the roots of the numerators or, alternatively, by multiplying each fraction by its respective numerator.

Let us consider several problems.

Problem 1

Prove the inequality for real numbers $a, b, c, d, e > 1$:

$$\frac{a^2}{c-1} + \frac{b^2}{d-1} + \frac{c^2}{e-1} + \frac{d^2}{a-1} + \frac{e^2}{b-1} \geq 20$$

Solution

Let us put $m = a + b + c + d + e$ and apply the Titu's Lemma to the numbers (a, b, c, d, e) and $(c-1, d-1, e-1, a-1, b-1)$:

$$\frac{a^2}{c-1} + \frac{b^2}{d-1} + \frac{c^2}{e-1} + \frac{d^2}{a-1} + \frac{e^2}{b-1} \geq \frac{(a+b+c+d+e)^2}{a+b+c+d+e-5} = \frac{m^2}{m-5}$$

It will be enough to prove that

$$\frac{m^2}{m-5} \geq 20$$

It is not hard to see that the last inequality holds since it is equivalent to an obvious inequality:

$$\frac{m^2}{m-5} \geq 20$$
$$m^2 \geq 20(m-5)$$
$$m^2 \geq 20m - 100$$
$$m^2 - 20m + 100 \geq 0$$
$$(m-10)^2 \geq 0$$

Problem 2

Prove that if a, b, c are the lengths of the sides of some triangle, then

$$\frac{a}{b+c-a} + \frac{b}{a+c-b} + \frac{c}{a+b-c} \geq 3$$

Solution

Let us start by rewriting the inequality as follows:

$$\frac{a^2}{ab+ac-a^2} + \frac{b^2}{ab+bc-b^2} + \frac{c^2}{ac+bc-c^2} \geq 3$$

Since the numerators of the fractions now contain squares, we can apply the Titu's Lemma to the numbers (a, b, c) and $\left(ab + ac - a^2, ab + bc - b^2, ac + bc - c^2\right)$:

$$\frac{a^2}{ab + ac - a^2} + \frac{b^2}{ab + bc - b^2} + \frac{c^2}{ac + bc - c^2} \geq \frac{(a + b + c)^2}{2ab + 2bc + 2ac - a^2 - b^2 - c^2}$$

It will be enough to prove that

$$\frac{(a + b + c)^2}{2ab + 2bc + 2ac - a^2 - b^2 - c^2} \geq 3$$

We will show that this inequality is equivalent to an obvious inequality. Let us multiply by the denominator and distribute:

$$(a + b + c)^2 \geq 3\left(2ab + 2bc + 2ac - a^2 - b^2 - c^2\right)$$

$$a^2 + b^2 + c^2 + 2ab + 2bc + 2ac \geq 6ab + 6bc + 6ac - 3a^2 - 3b^2 - 3c^2$$

Grouping the all the terms on the left and competing the squares we have

$$4a^2 + 4b^2 + 4c^2 - 4ab - 4bc - 4ac \geq 0$$

$$2\left(a^2 - 2ab + b^2\right) + 2\left(b^2 - 2bc + c^2\right) + 2\left(a^2 - 2ac + c^2\right) \geq 0$$

$$2(a - b)^2 + 2(b - c)^2 + 2(a - c)^2 \geq 0$$

which obviously holds.

Problem 3

Let a, b, c be positive real numbers. Prove that

$$\frac{a^2 + 2b^2}{2a + b} + \frac{b^2 + 2c^2}{2b + c} + \frac{c^2 + 2a^2}{2c + a} \geq a + b + c$$

Solution

Let us put

$$A = \frac{a^2}{2a + b} + \frac{b^2}{2b + c} + \frac{c^2}{2c + a}$$

$$B = \frac{b^2}{2a + b} + \frac{c^2}{2b + c} + \frac{a^2}{2c + a}$$

Then the left-hand side can be rewritten as follows:

$$\left(\frac{a^2}{2a + b} + \frac{b^2}{2b + c} + \frac{c^2}{2c + a}\right) + 2\left(\frac{b^2}{2a + b} + \frac{c^2}{2b + c} + \frac{a^2}{2c + a}\right) = A + 2B$$

It will be enough to prove that

$$A + 2B \geq a + b + c$$

By applying the Titu's Lemma to the numbers (a, b, c) and $(2a + b, 2b + c, 2c + a)$ we have that

$$A = \frac{a^2}{2a + b} + \frac{b^2}{2b + c} + \frac{c^2}{2c + a} \geq \frac{(a + b + c)^2}{3(a + b + c)} = \frac{a + b + c}{3}$$

By applying the Titu's Lemma to the tuples (b, c, a) and $(2a + b, 2b + c, 2c + a)$ we have that

$$B = \frac{b^2}{2a + b} + \frac{c^2}{2b + c} + \frac{a^2}{2c + a} \geq \frac{(a + b + c)^2}{3(a + b + c)} = \frac{a + b + c}{3}$$

Therefore, we have

$$A + 2B \geq \frac{a + b + c}{3} + \frac{2(a + b + c)}{3} = a + b + c$$

which is what needed to be proven.

CHAPTER 27

Bernoulli's Inequality

Bernoulli's Inequality

Bernoulli's Inequality states that for $x > -1$ and $n \geq 1$ the following inequality holds:
$$(1 + x)^n \geq 1 + nx$$
while for $x > -1$ and $0 < n \leq 1$ the following inequality holds:

$$(1 + x)^n \leq 1 + nx$$

The equality holds when $x = 0$ or $n = 1$.

Bernoulli's Inequality is commonly employed in problems involving expressions with exponents. To apply this inequality correctly, it is customary to represent the base as $1 + x$ if it is not immediately given in such a form.

Let us consider several problems.

Problem 1

Prove the inequality for the real numbers x_i from the interval $(0, 1]$:

$$(1 + x_1)^{\frac{1}{x_2}} \cdot (1 + x_2)^{\frac{1}{x_3}} \cdot \ldots \cdot (1 + x_n)^{\frac{1}{x_1}} \geq 2^n$$

Solution

Notice that since $x_i \leq 1$, then all the powers are $\frac{1}{x_i} \geq 1$. Now let us consider the first parenthesis and apply the Bernoulli's Inequality:

$$(1 + x_1)^{\frac{1}{x_2}} \geq 1 + \frac{x_1}{x_2}$$

However, by AM-GM Inequality we have

$$1 + \frac{x_1}{x_2} \geq 2\sqrt{\frac{x_1}{x_2}}$$

and then

$$(1 + x_1)^{\frac{1}{x_2}} \geq 1 + \frac{x_1}{x_2} \geq 2\sqrt{\frac{x_1}{x_2}}$$

Similarly, we can show that

$$(1 + x_2)^{\frac{1}{x_3}} \geq 1 + \frac{x_2}{x_3} \geq 2\sqrt{\frac{x_2}{x_3}}$$

$$\cdots \qquad \cdots \qquad \cdots$$

$$(1 + x_n)^{\frac{1}{x_1}} \geq 1 + \frac{x_n}{x_1} \geq 2\sqrt{\frac{x_n}{x_1}}$$

Multiplying these inequalities and telescoping the product under the root we have

$$(1 + x_1)^{\frac{1}{x_2}} \cdot (1 + x_2)^{\frac{1}{x_3}} \cdot \ldots \cdot (1 + x_n)^{\frac{1}{x_1}} \geq 2^n \sqrt{\frac{x_1}{x_2} \cdot \frac{x_2}{x_3} \cdot \ldots \cdot \frac{x_n}{x_1}} = 2^n$$

which is what needed to be proven.

Problem 2

Prove the inequality

$$x^y + y^z + z^x \leq xy + yz + zx + 2$$

for all positive x, y and z that satisfy the condition $x + y + z = 1$.

Solution

Start by noticing that all variables are less than 1. Let us consider the expression x^y and rewrite it as

$$x^y = (1 + (x - 1))^y$$

Let us now apply the Bernoulli's Inequality:

$$(1 + (x - 1))^y \leq 1 + (x - 1)y = 1 + xy - y$$

and, therefore, we have

$$x^y = (1 + (x - 1))^y \leq 1 + xy - y$$

Similarly, we have

$$y^z = (1 + (y - 1))^z \leq 1 + yz - z$$
$$z^x = (1 + (z - 1))^x \leq 1 + xz - x$$

Adding these inequalities we have

$$x^y + y^z + z^x \leq (1 + xy - y) + (1 + yz - z) + (1 + xz - x)$$
$$= 3 + xy + yz + xz - (x + y + z)$$
$$= xy + yz + xz + 2$$

as desired.

Problem 3

Let a, b, c be positive real numbers, such that $a + b + c < 1$. Prove that

$$a^{b+c} + b^{a+c} + c^{a+b} \geq 1$$

Solution

Start by noticing that $a + b < 1$, $b + c < 1$ and $a + c < 1$. Let us consider the following expression:

$$\left(\frac{1}{a}\right)^{b+c} = \left(1 + \frac{1-a}{a}\right)^{b+c}$$

and apply the Bernoulli's Inequality:

$$\left(1 + \frac{1-a}{a}\right)^{b+c} \leq 1 + \frac{(1-a)(b+c)}{a} = \frac{a+b+c-ab-ac}{a} < \frac{a+b+c}{a}$$

and, therefore, we have

$$\left(\frac{1}{a}\right)^{b+c} \le \frac{a+b+c}{a}$$

or equivalently

$$a^{b+c} \ge \frac{a}{a+b+c}$$

Similarly, we can show that

$$b^{a+c} \ge \frac{b}{a+b+c}$$

$$c^{a+b} \ge \frac{c}{a+b+c}$$

Adding these inequalities we have

$$a^{b+c} + b^{a+c} + c^{a+b} \ge \frac{a}{a+b+c} + \frac{b}{a+b+c} + \frac{c}{a+b+c} = 1$$

which is what needed to be proven.

CHAPTER 28

Jensen's Inequality and Concavity

Jensen's Inequality

Jensen's Inequality states that if the function $f(x)$ is concave upwards on the interval (a, b), then for the real numbers $x_1, x_2, \ldots, x_n$ from (a, b):

$$\frac{f(x_1) + f(x_2) + \ldots + f(x_n)}{n} \geq f\left(\frac{x_1 + x_2 + \ldots + x_n}{n}\right)$$

If the function $f(x)$ is concave downwards on the interval (a, b), then for the real numbers $x_1, x_2, \ldots, x_n$ from (a, b):

$$\frac{f(x_1) + f(x_2) + \ldots + f(x_n)}{n} \leq f\left(\frac{x_1 + x_2 + \ldots + x_n}{n}\right)$$

In this chapter we will focus on how to introduce the function $f(x)$, establish its concavity and then apply the **Jensen's Inequality**.

Concavity

The intervals where the function $f(x)$ is concave upwards or downwards are determined from the following:

- If the second derivative $f''(x)$ is positive on the interval (a, b), then $f(x)$ is concave upwards on (a, b).

- If the second derivative $f''(x)$ is negative on the interval (a, b), then $f(x)$ is concave downwards on (a, b).

Let us consider several problems.

Problem 1

If a, b, c are positive real numbers, prove that

$$\frac{a}{(b+c)^2} + \frac{b}{(c+a)^2} + \frac{c}{(a+b)^2} \geq \frac{9}{4(a+b+c)}$$

Solution

Let us make the substitution $t = a + b + c$. Then we have

$$b + c = t - a$$
$$c + a = t - b$$
$$a + b = t - c$$

and the inequality is equivalent to

$$\frac{a}{(t-a)^2} + \frac{b}{(t-b)^2} + \frac{c}{(t-c)^2} \geq \frac{9}{4t}$$

Notice that the function $f(x) = \frac{x}{(t-x)^2}$ is concave upwards on $(0, t)$. Indeed, it's second derivative is

$$f''(x) = \left(\frac{x}{(t-x)^2}\right)'' = \left(\frac{t+x}{(t-x)^3}\right)' = \frac{4t+2x}{(t-x)^4}$$

which is positive on $(0, t)$.

Since $a, b, c \in (0, t)$, then we can apply the Jensen's Inequality:

$$\frac{\frac{a}{(t-a)^2} + \frac{b}{(t-b)^2} + \frac{c}{(t-c)^2}}{3} \geq \frac{\frac{t}{3}}{\left(t - \frac{t}{3}\right)^2}$$

$$\frac{\frac{a}{(t-a)^2} + \frac{b}{(t-b)^2} + \frac{c}{(t-c)^2}}{3} \geq \frac{3}{4t}$$

$$\frac{a}{(t-a)^2} + \frac{b}{(t-b)^2} + \frac{c}{(t-c)^2} \geq \frac{9}{4t}$$

as desired.

Problem 2

Prove the inequality

$$\frac{1}{1+ab} + \frac{1}{1+bc} + \frac{1}{1+ac} \geq \frac{3}{2}$$

for positive a, b, c, such that $a^2 + b^2 + c^2 = 3$.

Solution

Start by noticing that the function $f(x) = \frac{1}{1+x}$ is concave upwards on $(0, \infty)$. Indeed, it's second derivative is

$$f''(x) = \left(\frac{1}{1+x}\right)'' = \left(-\frac{1}{(1+x)^2}\right)' = \frac{2}{(1+x)^3}$$

which is positive on $(0, \infty)$.

Since a, b and c are positive, then ab, bc and ac belong to the interval $(0, \infty)$ and by Jensen's Inequality we have:

$$\frac{\frac{1}{1+ab} + \frac{1}{1+bc} + \frac{1}{1+ac}}{3} \geq \frac{1}{1 + \frac{ab+bc+ac}{3}}$$

$$\frac{\frac{1}{1+ab} + \frac{1}{1+bc} + \frac{1}{1+ac}}{3} \geq \frac{3}{3 + ab + bc + ac}$$

$$\frac{1}{1+ab} + \frac{1}{1+bc} + \frac{1}{1+ac} \geq \frac{9}{3 + ab + bc + ac}$$

Now it will be enough to prove that

$$\frac{9}{3 + ab + bc + ac} \geq \frac{3}{2}$$

Let us rewrite this inequality as follows:

$$\frac{9}{3 + ab + bc + ac} \geq \frac{3}{2}$$

$$18 \geq 3(3 + ab + bc + ac)$$

$$18 \geq 9 + 3(ab + bc + ac)$$

$$9 \geq 3(ab + bc + ac)$$

$$3 \geq ab + bc + ac$$

Since $a^2 + b^2 + c^2 = 3$, then the last inequality is equivalent to

$$a^2 + b^2 + c^2 \geq ab + bc + ac$$

$$2a^2 + 2b^2 + 2c^2 \geq 2ab + 2bc + 2ac$$

$$\left(a^2 - 2ab + b^2\right) + \left(b^2 - 2bc + c^2\right) + \left(a^2 - 2ac + c^2\right) \geq 0$$

$$(a - b)^2 + (b - c)^2 + (a - c)^2 \geq 0$$

which obviously holds.

Problem 3

Prove that for positive $x_1, x_2, \ldots, x_n$, such that

$$x_1 + x_2 + \ldots + x_n = 1$$

the following inequality holds:

$$\sum_{i=1}^{n} x_i (1 - x_i)^2 \leq \left(\frac{n - 1}{n}\right)^2$$

Solution

We will solve this problem by doing the following casework:

- For $n = 1$ we have $x_1 = 1$ and the inequality obviously holds.
- For $n = 2$ we need to prove that

$$x_1 (1 - x_1)^2 + x_2 (1 - x_2)^2 \leq \left(\frac{1}{2}\right)^2$$

where $x_1 + x_2 = 1$. Squaring both sides of the last equation we have

$$(x_1 + x_2)^2 = (1)^2$$

$$x_1^2 + 2x_1 x_2 + x_2^2 = 1$$

$$x_1^2 + x_2^2 = 1 - 2x_1 x_2$$

Also from the AM-GM Inequality we have

$$x_1 + x_2 \geq 2\sqrt{x_1 x_2}$$
$$1 \geq 2\sqrt{x_1 x_2}$$
$$1 \geq 4x_1 x_2$$
$$\frac{1}{4} \geq x_1 x_2$$

Therefore, the initial inequality is equivalent to

$$x_1 (1 - x_1)^2 + x_2 (1 - x_2)^2 \leq \left(\frac{1}{2}\right)^2$$
$$x_1 + x_2 - 2x_1^2 - 2x_2^2 + x_1^3 + x_2^3 \leq \frac{1}{4}$$
$$1 - 2\left(x_1^2 + x_2^2\right) + (x_1 + x_2)\left(x_1^2 + x_2^2 - x_1 x_2\right) \leq \frac{1}{4}$$
$$1 - 2\left(x_1^2 + x_2^2\right) + \left(x_1^2 + x_2^2 - x_1 x_2\right) \leq \frac{1}{4}$$
$$1 - 2\left(1 - 2x_1 x_2\right) + \left(1 - 3x_1 x_2\right) \leq \frac{1}{4}$$
$$x_1 x_2 \leq \frac{1}{4}$$

which was shown before.

- For $n \geq 3$ start by noticing that the function $f(x) = x(1 - x)^2$ is concave upwards on $\left(-\infty, \frac{2}{3}\right)$. Indeed, it's second derivative is

$$f''(x) = \left(x(1-x)^2\right)'' = \left(x^3 - 2x^2 + x\right)'' = \left(3x^2 - 4x + 1\right)' = 6x - 4$$

which is positive on $\left(-\infty, \frac{2}{3}\right)$.

If all $x_i < \frac{2}{3}$, then by Jensen's Inequality we have:

$$\frac{\sum_{i=1}^{n} x_i (1 - x_i)^2}{n} \geq \frac{\sum_{i=1}^{n} x_i}{n} \cdot \left(1 - \frac{\sum_{i=1}^{n} x_i}{n}\right)^2$$
$$\frac{\sum_{i=1}^{n} x_i (1 - x_i)^2}{n} \geq \frac{1}{n} \cdot \left(1 - \frac{1}{n}\right)^2$$
$$\sum_{i=1}^{n} x_i (1 - x_i)^2 \geq \left(1 - \frac{1}{n}\right)^2$$
$$\sum_{i=1}^{n} x_i (1 - x_i)^2 \geq \left(\frac{n-1}{n}\right)^2$$

as desired.

If there exists $x_i \geq \frac{2}{3}$, then we have the following inequality:

$$x_i (1 - x_i)^2 \leq 1 \cdot \left(1 - \frac{2}{3}\right)^2 = \frac{1}{9}$$

For the rest of x_j we have

$$\sum_{\substack{j \neq i}}^{n} x_j (1 - x_j)^2 \leq \sum_{\substack{j \neq i}}^{n} x_j = 1 - x_i \leq \frac{1}{3}$$

Therefore, we have

$$\sum_{i=1}^{n} x_i (1 - x_i)^2 \leq \frac{1}{9} + \frac{1}{3} = \frac{4}{9} = \left(\frac{2}{3}\right)^2 \leq \left(\frac{n-1}{n}\right)^2$$

which is what needed to be proven.

CHAPTER 29

Jensen's Inequality in Geometric Inequalities

In this chapter we will look at several math olympiad problems that involve the concept of **Jensen's Inequality**. We will focus on the applications of the **Jensen's Inequality** to the geometric inequalities. We recommend reviewing the main ideas discussed in Chapter 28 "Jensen's Inequality and Concavity" before working on the exercises from this chapter.

Let us consider several problems.

Problem 1

Let R be the circumradius of the triangle ABC with the sides a, b and c. Prove that

$$a + b + c \leq 3\sqrt{3}R$$

Solution

Let us start by dividing both sides of the inequality by $2R$.

The inequality becomes:

$$\frac{a}{2R} + \frac{b}{2R} + \frac{c}{2R} \leq \frac{3\sqrt{3}}{2}$$

From the Law of Sines we have

$$\frac{a}{2R} = \sin A$$

$$\frac{b}{2R} = \sin B$$

$$\frac{c}{2R} = \sin C$$

and, therefore, the inequality is equivalent to

$$\sin A + \sin B + \sin C \leq \frac{3\sqrt{3}}{2}$$

Notice that the function $f(x) = \sin x$ is concave downwards on $(0, \pi)$. Indeed, it's second derivative is

$$f''(x) = (\sin x)'' = (\cos x)' = -\sin x$$

which is negative on $(0, \pi)$.

Since A, B and C are the angles of a triangle, then they belong to the interval $(0, \pi)$ and the last inequality directly follows from the Jensen's Inequality:

$$\frac{\sin A + \sin B + \sin C}{3} \leq \sin\left(\frac{A + B + C}{3}\right)$$

$$\frac{\sin A + \sin B + \sin C}{3} \leq \sin\left(\frac{\pi}{3}\right)$$

$$\frac{\sin A + \sin B + \sin C}{3} \leq \frac{\sqrt{3}}{2}$$

$$\sin A + \sin B + \sin C \leq \frac{3\sqrt{3}}{2}$$

as desired.

Problem 2

Let a, b, c be the sides of an acute triangle ABC. Prove that

$$\sum_{\text{cyc}} \frac{a^2 + b^2 - c^2}{ab} \leq 3$$

Solution

Start by noticing that by the Law of Cosines

$$c^2 = a^2 + b^2 - 2ab\cos C$$

$$2ab\cos C = a^2 + b^2 - c^2$$

$$2\cos C = \frac{a^2 + b^2 - c^2}{ab}$$

and, therefore, the inequality becomes

$$2\cos A + 2\cos B + 2\cos C \le 3$$

or equivalently

$$\cos A + \cos B + \cos C \le \frac{3}{2}$$

Notice that the function $f(x) = \cos x$ is concave downwards on $\left(0, \frac{\pi}{2}\right)$. Indeed, it's second derivative is

$$f''(x) = (\cos x)'' = (-\sin x)' = -\cos x$$

which is negative on $\left(0, \frac{\pi}{2}\right)$.

Since A, B and C are the angles of an acute triangle, then they belong to the interval $\left(0, \frac{\pi}{2}\right)$ and the last inequality directly follows from the Jensen's Inequality:

$$\frac{\cos A + \cos B + \cos C}{3} \le \cos\left(\frac{A + B + C}{3}\right)$$

$$\frac{\cos A + \cos B + \cos C}{3} \le \cos\left(\frac{\pi}{3}\right)$$

$$\frac{\cos A + \cos B + \cos C}{3} \le \frac{1}{2}$$

$$\cos A + \cos B + \cos C \le \frac{3}{2}$$

as desired.

Problem 3

Let h_a, h_b, h_c be the altitudes drawn from the vertices A, B, C of an acute triangle ABC. Prove that

$$\frac{a}{h_a} + \frac{b}{h_b} + \frac{c}{h_c} \ge 2\sqrt{3}$$

Solution

Let A_1, B_1, C_1 be the feet of the altitudes h_a, h_b, h_c respectively.

Start by noticing that

$$\cot A = \frac{AB_1}{h_b}$$

$$\cot C = \frac{CB_1}{h_b}$$

and, therefore

$$\cot A + \cot C = \frac{AB_1}{h_b} + \frac{CB_1}{h_b} = \frac{AB_1 + CB_1}{h_b} = \frac{AC}{h_b} = \frac{b}{h_b}$$

Similarly, we can show that

$$\cot A + \cot B = \frac{c}{h_c}$$

$$\cot B + \cot C = \frac{a}{h_a}$$

Then the inequality becomes

$$2\left(\cot A + \cot B + \cot C\right) \geq 2\sqrt{3}$$

which is equivalent to

$$\cot A + \cot B + \cot C \geq \sqrt{3}$$

Notice that the function $f(x) = \cot x$ is concave upwards on $\left(0, \frac{\pi}{2}\right)$. Indeed, it's second derivative is

$$f''(x) = (\cot x)'' = \left(-\csc^2 x\right)' = 2\cot x \csc^2 x$$

which is positive on $\left(0, \frac{\pi}{2}\right)$.

Since A, B and C are the angles of an acute triangle, then they belong to the interval $\left(0, \frac{\pi}{2}\right)$ and the last inequality directly follows from the Jensen's Inequality:

$$\frac{\cot A + \cot B + \cot C}{3} \geq \cot\left(\frac{A+B+C}{3}\right)$$

$$\frac{\cot A + \cot B + \cot C}{3} \geq \cot\left(\frac{\pi}{3}\right)$$

$$\frac{\cot A + \cot B + \cot C}{3} \geq \frac{\sqrt{3}}{3}$$

$$\cot A + \cot B + \cot C \geq \sqrt{3}$$

as desired.

CHAPTER 30

Rearrangement Inequality

Rearrangement Inequality

Given real numbers $(a_1, a_2, \ldots, a_n)$ and $(b_1, b_2, \ldots, b_n)$, such that $a_1 \leq a_2 \leq \ldots \leq a_n$ and $b_1 \leq b_2 \leq \ldots \leq b_n$. Let $(c_1, c_2, \ldots, c_n)$ be some permutation of $(b_1, b_2, \ldots, b_n)$. **Rearrangement Inequality** states that

$$S \geq P \geq R$$

where **sorted sum** S, **reversed sum** R and **permuted sum** P are defined as

$$S = a_1 b_1 + a_2 b_2 + \ldots + a_n b_n$$
$$R = a_1 b_n + a_2 b_{n-1} + \ldots + a_n b_1$$
$$P = a_1 c_1 + a_2 c_2 + \ldots + a_n c_n$$

Let us consider several problems.

Problem 1

Prove the inequality for positive a, b, c

$$a^4 + b^4 + c^4 \geq abc(a + b + c)$$

Solution

Notice that the right-hand side of the inequality can be written as

$$abc(a + b + c) = a^2bc + ab^2c + abc^2 = \sum_{cyc} a^2bc$$

Let us apply the Rearrangement Inequality twice to the numbers (a^2, b^2, c^2) and (a^2, b^2, c^2).

Permuting once we have

$$a^4 + b^4 + c^4 \geq a^2b^2 + b^2c^2 + c^2a^2 = \sum_{cyc} a^2b^2$$

Permuting twice we have

$$a^4 + b^4 + c^4 \geq a^2c^2 + b^2a^2 + c^2b^2 = \sum_{cyc} a^2c^2$$

Adding these inequalities and applying the AM-GM Inequality we have

$$2\left(a^4 + b^4 + c^4\right) \geq \sum_{cyc} a^2b^2 + \sum_{cyc} a^2c^2$$
$$= \sum_{cyc} a^2\left(b^2 + c^2\right)$$
$$\geq \sum_{cyc} a^2(2bc)$$
$$= 2\sum_{cyc} a^2bc$$

and, therefore

$$a^4 + b^4 + c^4 \geq abc(a + b + c)$$

as desired.

Problem 2

Prove the inequality for positive a, b, c, such that $abc = 1$:

$$\frac{a^2}{b+c} + \frac{b^2}{c+a} + \frac{c^2}{a+b} \geq \frac{3}{2}$$

Solution

Let us apply the Rearrangement Inequality twice to the numbers $\left(a^2, b^2, c^2\right)$ and $\left(\frac{1}{b+c}, \frac{1}{c+a}, \frac{1}{a+b}\right)$.

Permuting once we have

$$\frac{a^2}{b+c} + \frac{b^2}{c+a} + \frac{c^2}{a+b} \geq \frac{a^2}{a+b} + \frac{b^2}{b+c} + \frac{c^2}{c+a} = \sum_{\text{cyc}} \frac{a^2}{a+b}$$

Permuting twice we have

$$\frac{a^2}{b+c} + \frac{b^2}{c+a} + \frac{c^2}{a+b} \geq \frac{a^2}{c+a} + \frac{b^2}{a+b} + \frac{c^2}{b+c} = \sum_{\text{cyc}} \frac{b^2}{a+b}$$

Adding these inequalities we have

$$2\left(\frac{a^2}{b+c} + \frac{b^2}{c+a} + \frac{c^2}{a+b}\right) \geq \sum_{\text{cyc}} \frac{a^2}{a+b} + \sum_{\text{cyc}} \frac{b^2}{a+b}$$

$$= \sum_{\text{cyc}} \frac{a^2 + b^2}{a+b}$$

$$\geq \sum_{\text{cyc}} \frac{a+b}{2}$$

$$= a + b + c$$

Now by AM-GM Inequality we have

$$a + b + c \geq 3\sqrt[3]{abc} = 3$$

and, therefore, we have

$$\frac{a^2}{b+c} + \frac{b^2}{c+a} + \frac{c^2}{a+b} \geq \frac{3}{2}$$

which is what needed to be proven.

Problem 3

Consider three positive real numbers x, y, z whose product is 1. Prove that

$$\frac{xy}{x^5 + xy + y^5} + \frac{yz}{y^5 + yz + z^5} + \frac{zx}{z^5 + zx + x^5} \le 1$$

Solution

Notice that the left-hand side of the inequality can be written as

$$\frac{xy}{x^5 + xy + y^5} + \frac{yz}{y^5 + yz + z^5} + \frac{zx}{z^5 + zx + x^5} = \sum_{\text{cyc}} \frac{xy}{x^5 + xy + y^5}$$

Let us apply the Rearrangement Inequality to the numbers $\left(x^3, y^3\right)$ and $\left(x^2, y^2\right)$:

$$x^5 + y^5 \ge x^3 y^2 + x^2 y^3$$

Therefore, we have

$$\sum_{\text{cyc}} \frac{xy}{x^5 + xy + y^5} \le \sum_{\text{cyc}} \frac{xy}{x^3 y^2 + xy + x^2 y^3}$$

$$= \sum_{\text{cyc}} \frac{xyz^2}{x^3 y^2 z^2 + xyz^2 + x^2 y^3 z^2}$$

$$= \sum_{\text{cyc}} \frac{(xyz) \cdot z}{(xyz)^2 \cdot x + (xyz) \cdot y + (xyz)^2 \cdot z}$$

$$= \sum_{\text{cyc}} \frac{z}{x + y + z}$$

$$= \frac{z + x + y}{x + y + z}$$

$$= 1$$

which is what needed to be proven.

CHAPTER 31

Chebyshev's Inequality

Chebyshev's Inequality

Given real numbers $(a_1, a_2, \ldots, a_n)$ and $(b_1, b_2, \ldots, b_n)$, such that $a_1 \leq a_2 \leq \ldots \leq a_n$ and $b_1 \leq b_2 \leq \ldots \leq b_n$. **Chebyshev's Inequality** states that

$$S \geq A \geq R$$

where **sorted sum** S, **reversed sum** R and **averaged product** A are defined as follows:

$$S = a_1 b_1 + a_2 b_2 + \ldots + a_n b_n$$
$$R = a_1 b_n + a_2 b_{n-1} + \ldots + a_n b_1$$
$$A = \frac{1}{n}(a_1 + a_2 + \ldots + a_n)(b_1 + b_2 + \ldots + b_n)$$

Let us consider several problems.

Problem 1

Prove the inequality for positive values of the variables

$$a^a b^b c^c \geq (abc)^d$$

where $a + b + c = 3d$.

Solution

Let us apply the Chebyshev's Inequality to the numbers (a, b, c) and $(\ln a, \ln b, \ln c)$:

$$a \ln a + b \ln b + c \ln c \geq \frac{1}{3}(a + b + c)(\ln a + \ln b + \ln c)$$

$$\ln a^a + \ln b^b + \ln c^c \geq d \ln abc$$

$$\ln a^a + \ln b^b + \ln c^c \geq \ln (abc)^d$$

$$\ln a^a b^b c^c \geq \ln (abc)^d$$

$$a^a b^b c^c \geq (abc)^d$$

which is what needed to be proven.

Problem 2

Prove that for $a, b, c > 0$, the following inequality holds

$$\frac{a^8 + b^8 + c^8}{(abc)^3} \geq \frac{1}{a} + \frac{1}{b} + \frac{1}{c}$$

Solution

Let us apply the Chebyshev's Inequality to the numbers $\left(a^6, b^6, c^6\right)$ and $\left(a^2, b^2, c^2\right)$:

$$a^8 + b^8 + c^8 \geq \frac{1}{3}\left(a^6 + b^6 + c^6\right)\left(a^2 + b^2 + c^2\right)$$

However, by the AM-GM Inequality we have

$$a^6 + b^6 + c^6 \geq 3\sqrt[3]{a^6 b^6 c^6} = 3a^2 b^2 c^2$$

and also

$$a^2 + b^2 + c^2 \geq ab + bc + ca$$

Therefore, we have shown that

$$a^8 + b^8 + c^8 \geq a^2 b^2 c^2 (ab + bc + ca)$$

From here we have

$$\frac{a^8 + b^8 + c^8}{(abc)^3} \geq \frac{a^2 b^2 c^2 \left(ab + bc + ca\right)}{(abc)^3} = \frac{ab + bc + ca}{abc} = \frac{1}{a} + \frac{1}{b} + \frac{1}{c}$$

which is what needed to be proven.

Problem 3

Prove that if $a, b, c > 0$, then

$$\frac{a^3}{bc} + \frac{b^3}{ca} + \frac{c^3}{ab} \geq a + b + c$$

Solution

Let us apply the Chebyshev's Inequality to the numbers $\left(a^3, b^3, c^3\right)$ and $\left(\frac{1}{bc}, \frac{1}{ca}, \frac{1}{ab}\right)$:

$$\frac{a^3}{bc} + \frac{b^3}{ca} + \frac{c^3}{ab} \geq \frac{1}{3} \left(a^3 + b^3 + c^3\right) \left(\frac{1}{bc} + \frac{1}{ca} + \frac{1}{ab}\right)$$

However, by Titu's Lemma we have

$$\frac{1}{bc} + \frac{1}{ca} + \frac{1}{ab} \geq \frac{(1 + 1 + 1)^2}{ab + bc + ca} = \frac{9}{ab + bc + ca}$$

Therefore, we have shown that

$$\frac{a^3}{bc} + \frac{b^3}{ca} + \frac{c^3}{ab} \geq \frac{3 \left(a^3 + b^3 + c^3\right)}{ab + bc + ca}$$

Let us apply the Chebyshev's Inequality to the numbers (a, b, c) and $\left(a^2, b^2, c^2\right)$:

$$a^3 + b^3 + c^3 \geq \frac{1}{3} (a + b + c) \left(a^2 + b^2 + c^2\right)$$

Since $a^2 + b^2 + c^2 \geq ab + bc + ca$, we have

$$\frac{a^3}{bc} + \frac{b^3}{ca} + \frac{c^3}{ab} \geq \frac{(a + b + c) \left(a^2 + b^2 + c^2\right)}{ab + bc + ca} \geq a + b + c$$

which is what needed to be proven.

CHAPTER 32

Hölder's Inequality

Hölder's Inequality

For the tuples of positive real numbers

$$(a_{11}, a_{12}, \ldots, a_{1n})$$
$$(a_{21}, a_{22}, \ldots, a_{2n})$$
$$\vdots \qquad \qquad \vdots$$
$$(a_{m1}, a_{m2}, \ldots, a_{mn})$$

the following inequality holds:

$$\prod_{i=1}^{m} \left(\sum_{j=1}^{n} a_{ij} \right) \geq \left(\sum_{j=1}^{n} \sqrt[m]{\prod_{i=1}^{m} a_{ij}} \right)^{m}$$

Let us consider several problems.

Problem 1

For the positive real numbers a, b, c, prove that

$$\left(3 + a^4\right)\left(3 + b^4\right)\left(3 + c^4\right)\left(3 + d^4\right) \geq (a + b + c + d)^4$$

Solution

Let us apply Hölder's Inequality to the four tuples:

$$\left(a^4, 1, 1, 1\right)$$
$$\left(1, b^4, 1, 1\right)$$
$$\left(1, 1, c^4, 1\right)$$
$$\left(1, 1, 1, d^4\right)$$

The left-hand side of Hölder's Inequality becomes

$$\text{LHS} = \left(a^4 + 1 + 1 + 1\right)\left(b^4 + 1 + 1 + 1\right)\left(c^4 + 1 + 1 + 1\right)\left(d^4 + 1 + 1 + 1\right)$$
$$= \left(a^4 + 3\right)\left(b^4 + 3\right)\left(c^4 + 3\right)\left(d^4 + 3\right)$$

while the right-hand side of Hölder's Inequality becomes

$$\text{RHS} = \left(\sqrt[4]{a^4 \cdot 1 \cdot 1 \cdot 1} + \sqrt[4]{b^4 \cdot 1 \cdot 1 \cdot 1} + \sqrt[4]{c^4 \cdot 1 \cdot 1 \cdot 1} + \sqrt[4]{d^4 \cdot 1 \cdot 1 \cdot 1}\right)^4$$
$$= \left(\sqrt[4]{a^4} + \sqrt[4]{b^4} + \sqrt[4]{c^4} + \sqrt[4]{d^4}\right)^4$$
$$= (a + b + c + d)^4$$

From here

$$\left(3 + a^4\right)\left(3 + b^4\right)\left(3 + c^4\right)\left(3 + d^4\right) \geq (a + b + c + d)^4$$

as desired.

Problem 2

Prove the inequality for positive numbers

$$\left(a^2 + ab + b^2\right)\left(b^2 + bc + c^2\right)\left(c^2 + ca + a^2\right) \geq (ab + bc + ca)^3$$

Solution

Let us apply Hölder's Inequality to the three tuples:

$$\left(ab, a^2, b^2\right)$$
$$\left(b^2, c^2, bc\right)$$
$$\left(a^2, ac, c^2\right)$$

The left-hand side of Hölder's Inequality becomes

$$\begin{aligned}
\text{LHS} &= \left(ab + a^2 + b^2\right)\left(b^2 + c^2 + bc\right)\left(a^2 + ac + c^2\right) \\
&= \left(a^2 + ab + b^2\right)\left(b^2 + bc + c^2\right)\left(c^2 + ca + a^2\right)
\end{aligned}$$

while the right-hand side of Hölder's Inequality becomes

$$\begin{aligned}
\text{RHS} &= \left(\sqrt[3]{ab \cdot b^2 \cdot a^2} + \sqrt[3]{a^2 \cdot c^2 \cdot ac} + \sqrt[3]{b^2 \cdot bc \cdot c^2}\right)^3 \\
&= \left(\sqrt[3]{(ab)^3} + \sqrt[3]{(ca)^3} + \sqrt[3]{(bc)^3}\right)^3 \\
&= (ab + bc + ca)^3
\end{aligned}$$

From here

$$\left(a^2 + ab + b^2\right)\left(b^2 + bc + c^2\right)\left(c^2 + ca + a^2\right) \geq (ab + bc + ca)^3$$

as desired.

Problem 3

Let a, b, c be three positive real numbers, such that $a + b + c = 1$. Prove that

$$\sum_{\text{cyc}} \frac{a}{\sqrt[3]{a + 2b}} \geq 1$$

Solution

Let us apply Hölder's Inequality to the four tuples:

$$\left(\frac{a}{\sqrt[3]{a + 2b}}, \frac{b}{\sqrt[3]{b + 2c}}, \frac{c}{\sqrt[3]{c + 2a}}\right)$$

$$\left(\frac{a}{\sqrt[3]{a + 2b}}, \frac{b}{\sqrt[3]{b + 2c}}, \frac{c}{\sqrt[3]{c + 2a}}\right)$$

$$\left(\frac{a}{\sqrt[3]{a + 2b}}, \frac{b}{\sqrt[3]{b + 2c}}, \frac{c}{\sqrt[3]{c + 2a}}\right)$$

$$(a(a + 2b), b(b + 2c), c(c + 2a))$$

Notice that

$$\begin{aligned}
a(a + 2b) + b(b + 2c) + c(c + 2a) &= a^2 + 2ab + b^2 + 2bc + c^2 + ac \\
&= (a + b + c)^2 \\
&= (1)^2 \\
&= 1
\end{aligned}$$

The left-hand side of Hölder's Inequality becomes

$$\begin{aligned}
\text{LHS} &= \left(\frac{a}{\sqrt[3]{a + 2b}} + \frac{b}{\sqrt[3]{b + 2c}} + \frac{c}{\sqrt[3]{c + 2a}} \right)^3 \\
&= \left(\sum_{\text{cyc}} \frac{a}{\sqrt[3]{a + 2b}} \right)^3
\end{aligned}$$

while the right-hand side of Hölder's Inequality becomes

$$\begin{aligned}
\text{RHS} &= \left(\sqrt[4]{\frac{a^3}{a + 2b}} \cdot a(a + 2b) + \sqrt[4]{\frac{b^3}{b + 2c}} \cdot b(b + 2c) + \sqrt[4]{\frac{c^3}{c + 2a}} \cdot c(a + 2a) \right)^4 \\
&= \left(\sqrt[4]{a^4} + \sqrt[4]{b^4} + \sqrt[4]{c^4} \right)^4 \\
&= (a + b + c)^4 \\
&= (1)^4 \\
&= 1
\end{aligned}$$

From here

$$\sum_{\text{cyc}} \frac{a}{\sqrt[3]{a + 2b}} \geq 1$$

as desired.

CHAPTER 33

Tangent Line Technique

Tangent Line Technique

Tangent Line Technique relies on using one of the fundamental inequalities

$$f(x) \geq t(x) \quad \text{or} \quad f(x) \leq t(x)$$

where $f(x)$ is some function and $t(x)$ is the equation of its tangent line drawn at some point. Graphically this means that the function $f(x)$ on some interval lies either completely above or completely below its tangent line.

The needed fundamental inequality can be proven either from the concavity of the function $f(x)$ (concave upwards functions lie above their tangent line while concave downwards functions lie below their tangent line) or can be proven by transforming the inequality to an obvious inequality. The fundamental inequality is then used for each of the variables of the original inequality. The combination of the obtained inequalities usually implies the inequality of the problem.

Let us consider several problems.

Problem 1

Let a, b, c be positive numbers, such that $a + b + c = 3$. Prove

$$a^3b + b^3c + c^3a + 6 \geq 3(ab + bc + ac)$$

Solution

Start by noticing that since the variables are positive and $a + b + c = 3$, then we have that $a, b, c \in (0, 3)$.

Let us consider the function $f(x) = x^2$ and find its tangent line at the point $(1, 1)$. The derivative of the function $f(x)$ is

$$f'(x) = \left(x^3\right)' = 3x^2$$

The slope of the tangent line m is equal to

$$m = f'(1) = 3(1)^2 = 3$$

From the Point-Slope Formula we can obtain the equation of the tangent line:

$$y - y_1 = m(x - x_1)$$
$$y - 1 = 3(x - 1)$$
$$y - 1 = 3x - 3$$
$$y = 3x - 2$$

The second derivative of the function $f(x)$ is

$$f''(x) = \left(3x^2\right)' = 6x$$

which is positive on $(0, 3)$. This implies that the function $f(x)$ is concave upwards on $(0, 3)$ and its tangent line lies below the function $f(x)$. Therefore, we have the following fundamental inequality:

$$x^3 \geq 3x - 2$$

Let us apply the fundamental inequality to the variable a and multiply it by b:

$$a^3 \geq 3a - 2$$
$$a^3b \geq 3ab - 2b$$
$$a^3b + 2b \geq 3ab$$

Similarly, we can show that

$$b^3c + 2c \geq 3bc$$
$$c^3a + 2a \geq 3ac$$

Adding these inequalities and taking into account that $a + b + c = 3$ we have

$$a^3b + b^3c + c^3a + 2(a + b + c) \geq 3(ab + bc + ac)$$
$$a^3b + b^3c + c^3a + 6 \geq 3(ab + bc + ac)$$

which is what needed to be proven.

Problem 2

Let a, b, c, d be positive real numbers, such that $a + b + c + d = 1$. Prove that

$$8\left(a^3 + b^3 + c^3 + d^3\right) \geq a^2 + b^2 + c^2 + d^2 + \frac{1}{4}$$

Solution

Start by noticing that since the variables are positive and $a + b + c + d = 1$, then we have that $a, b, c, d \in (0, 1)$.

Let us consider the function $f(x) = 8x^3 - x^2$ and find its tangent line at the point $\left(\frac{1}{4}, \frac{1}{16}\right)$. The derivative of the function $f(x)$ is

$$f'(x) = \left(8x^3 - x^2\right)' = 24x^2 - 2x$$

The slope of the tangent line m is equal to

$$m = f'\left(\frac{1}{4}\right) = 24\left(\frac{1}{4}\right)^2 - 2\left(\frac{1}{4}\right) = 1$$

From the Point-Slope Formula we can obtain the equation of the tangent line:

$$y - y_1 = m\left(x - x_1\right)$$
$$y - \frac{1}{16} = 1\left(x - \frac{1}{4}\right)$$
$$y - \frac{1}{16} = x - \frac{1}{4}$$
$$y = x - \frac{3}{16}$$

This suggests the following fundamental inequality:

$$8x^3 - x^2 \geq x - \frac{3}{16}$$

Unfortunately, in this case the function $f(x)$ changes its concavity and we will not be able to use the second derivative. However, the inequality still holds and we can prove it by transforming it to an obvious inequality:

$$8x^3 - x^2 \geq x - \frac{3}{16}$$
$$128x^3 - 16x^2 \geq 16x - 3$$
$$128x^3 - 16x^2 - 16x + 3 \geq 0$$
$$(8x + 3)(4x - 1)^2 \geq 0$$

Let us now apply the fundamental inequality to the variables a, b, c and d:

$$8a^3 - a^2 \geq a - \frac{3}{16}$$
$$8b^3 - b^2 \geq b - \frac{3}{16}$$
$$8c^3 - c^2 \geq c - \frac{3}{16}$$
$$8d^3 - d^2 \geq d - \frac{3}{16}$$

Adding these inequalities and taking into account that $a + b + c + d = 1$ we have

$$8\left(a^3 + b^3 + c^3 + d^3\right) - \left(a^2 + b^2 + c^2 + d^2\right) \geq (a + b + c + d) - 4\left(\frac{3}{16}\right)$$
$$8\left(a^3 + b^3 + c^3 + d^3\right) - \left(a^2 + b^2 + c^2 + d^2\right) \geq \frac{1}{4}$$
$$8\left(a^3 + b^3 + c^3 + d^3\right) \geq a^2 + b^2 + c^2 + d^2 + \frac{1}{4}$$

which is what needed to be proven.

Problem 3

Given nonnegative real numbers a, b, c, d, such that $a + b + c = 6$. Prove the inequality

$$\frac{a}{b^3 + 4} + \frac{b}{c^3 + 4} + \frac{c}{a^3 + 4} \geq \frac{1}{2}$$

Solution

Let us consider the function $f(x) = \frac{1}{x^3+4}$ and find its tangent line at the point $\left(2, \frac{1}{12}\right)$.

The derivative of the function $f(x)$ is

$$f'(x) = \left(\frac{1}{x^3+4}\right)' = -\frac{3x^2}{(x^3+4)^2}$$

The slope of the tangent line m is equal to

$$m = f'(2) = -\frac{3(2)^2}{((2)^3+4)^2} = -\frac{1}{12}$$

From the Point-Slope Formula we can obtain the equation of the tangent line:

$$y - y_1 = m(x - x_1)$$

$$y - \frac{1}{12} = -\frac{1}{12}(x-2)$$

$$y - \frac{1}{12} = -\frac{x}{12} + \frac{1}{6}$$

$$y = -\frac{x}{12} + \frac{1}{4}$$

This suggests the following fundamental inequality:

$$\frac{1}{x^3+4} \geq -\frac{x}{12} + \frac{1}{4}$$

Unfortunately, in this case the function $f(x)$ changes its concavity and we will not be able to use the second derivative. However, the inequality still holds and we can prove it by transforming it to an obvious inequality:

$$\frac{1}{x^3+4} \geq -\frac{x}{12} + \frac{1}{4}$$

$$\frac{12}{x^3+4} \geq -x + 3$$

$$12 \geq (x^3+4)(3-x)$$

$$12 \geq -x^4 + 3x^3 - 4x + 12$$

$$x^4 - 3x^3 + 4x \geq 0$$

$$x(x+1)(x-2)^2 \geq 0$$

Let us apply the fundamental inequality to the variable b and multiply it by a:

$$\frac{1}{b^3+4} \geq -\frac{b}{12} + \frac{1}{4}$$

$$\frac{a}{b^3+4} \geq -\frac{ab}{12} + \frac{a}{4}$$

Similarly, we can show that

$$\frac{b}{c^3 + 4} \geq -\frac{bc}{12} + \frac{b}{4}$$

$$\frac{c}{a^3 + 4} \geq -\frac{ca}{12} + \frac{c}{4}$$

Adding these inequalities and taking into account that $a + b + c = 6$ we have

$$\frac{a}{b^3 + 4} + \frac{b}{c^3 + 4} + \frac{c}{a^3 + 4} \geq -\frac{ab + bc + ca}{12} + \frac{a + b + c}{4}$$

However, since

$$6^2 = (a + b + c)^2 = a^2 + b^2 + c^2 + 2(ab + bc + ca) \geq 3(ab + bc + ca)$$

then

$$ab + bc + ca \leq 12$$

Therefore, we have

$$\frac{a}{b^3 + 4} + \frac{b}{c^3 + 4} + \frac{c}{a^3 + 4} \geq -\frac{12}{12} + \frac{6}{4} = \frac{1}{2}$$

which is what needed to be proven.

CHAPTER 34

Elimination of Variables

Elimination of Variables is a technique of solving math olympiad problems by converting them to problems with only one variable and then using the derivative to investigate the maximum/minimum of the function and the intervals where the function is increasing/decreasing.

Let us consider several problems.

Problem 1

Given three real numbers, such that their sum is 2 and the sum of their squares is 4. Find the minimum value of their product.

Solution

Let the numbers in the problem be a, b and c.

Therefore, we are given that

$$a + b + c = 2$$
$$a^2 + b^2 + c^2 = 4$$

We will show that the minimum value of abc is $-\frac{32}{27}$.

Let us start by showing that $a, b, c \in \left[-\frac{2}{3}, 2\right]$. Let us multiply the second equation by 2 and rewrite it as follows:

$$
\begin{aligned}
0 &= 2\left(a^2 - 4\right) + 2\left(b^2 + c^2\right) \\
&\geq 2\left(a^2 - 4\right) + (b + c)^2 \\
&= 2\left(a^2 - 4\right) + (2 - a)^2 \\
&= 2\left(a^2 - 4\right) + \left(a^2 - 4a + 4\right) \\
&= 3a^2 - 4a - 4 \\
&= (3a + 2)(a - 2)
\end{aligned}
$$

From here $a \in \left[-\frac{2}{3}, 2\right]$. Similarly, it can be shown that $b, c \in \left[-\frac{2}{3}, 2\right]$.

Let us now find the value of $ab + bc + ac$ by squaring both sides of the first equality:

$$
\begin{aligned}
(a + b + c)^2 &= (2)^2 \\
a^2 + b^2 + c^2 + 2(ab + bc + ac) &= 4 \\
4 + 2(ab + bc + ac) &= 4 \\
2(ab + bc + ac) &= 0 \\
ab + bc + ac &= 0
\end{aligned}
$$

Let us consider the polynomial $P(x)$ defined as

$$P(x) = (x - a)(x - b)(x - c)$$

which has its roots at $x = a$, $x = b$ and $x = c$.

From the Vieta's Formulas[1] we have

$$P(x) = x^3 - (a + b + c)x^2 + (ab + bc + ac)x - abc$$

and the polynomial $P(x)$ becomes

$$P(x) = x^3 - 2x^2 - abc$$

---

[1] These formulas are discussed in detail in Chapter 14 "Vieta's Formulas for Cubic Polynomial".

The equality $P(x) = 0$ holds for the values of a, b and c and, therefore, we have

$$x^3 - 2x^2 - abc = 0$$

from where

$$abc = x^3 - 2x^2$$

Let us consider the function $f(t) = t^3 - 2t^2$ on the interval $\left[-\frac{2}{3}, 2\right]$.

Its first derivative is

$$f'(t) = \left(t^3 - 2t^2\right)' = 3t^2 - 4t$$

and the critical points of $f(t)$ are $t = 0$ and $t = \frac{4}{3}$.

The derivative $f'(t)$ is positive on the interval $\left(-\frac{2}{3}, 0\right)$ and, therefore, the function $f(t)$ increases on $\left[-\frac{2}{3}, 0\right]$. The derivative $f'(t)$ is negative on the interval $\left(0, \frac{4}{3}\right)$ and, therefore, the function $f(t)$ decreases on $\left[0, \frac{4}{3}\right]$. The derivative $f'(t)$ is positive on the interval $\left(\frac{4}{3}, 2\right)$ and, therefore, the function $f(t)$ increases on $\left[\frac{4}{3}, 2\right]$.

This implies that $f(t) \geq f\left(\frac{4}{3}\right)$ for $t \in [0, 2]$, and $f(t) \geq f\left(-\frac{2}{3}\right)$ for $t \in \left[-\frac{2}{3}, 0\right]$.

However, since

$$f\left(\frac{4}{3}\right) = f\left(-\frac{2}{3}\right) = -\frac{32}{27}$$

we have that

$$f(t) \geq -\frac{32}{27}$$

for all $t \in \left[-\frac{2}{3}, 2\right]$.

It is not hard to see that the equality is reached, for example, when $a = -\frac{2}{3}$, $b = \frac{4}{3}$, $c = \frac{4}{3}$.

Problem 2

Solve the equation in positive real numbers

$$a^3 + b^3 + c^3 - a^2b - b^2c - c^2a = 0$$

Solution

Without loss of generality we can assume that $a \geq b \geq c$. Let us divide both sides of the equation by b^3 and transform it to an equivalent equation as follows:

$$\frac{a^3}{b^3} + \frac{b^3}{b^3} + \frac{c^3}{b^3} - \frac{a^2b}{b^3} - \frac{b^2c}{b^3} - \frac{c^2a}{b^3} = \frac{0}{b^3}$$

$$\left(\frac{a}{b}\right)^3 + 1 + \left(\frac{c}{b}\right)^3 - \left(\frac{a}{b}\right)^2 - \left(\frac{c}{b}\right) - \left(\frac{c}{b}\right)^2 \left(\frac{a}{b}\right) = 0$$

Let us now make the substitutions $x = \frac{a}{b}$ and $y = \frac{c}{b}$, where x and y are positive real numbers. Notice that since $a \geq b$, then $x \geq 1$, and since $b \geq c$, then $y \leq 1$. The equation now becomes

$$x^3 + 1 + y^3 - x^2 - y - y^2 x = 0$$

Let us consider the function

$$f(t) = t^3 - t^2 - y^2 t + y^3 - y + 1$$

Its first derivative is

$$f'(t) = 3t^2 - 2t - y^2$$

Since $0 < y \leq 1$, then $0 < y^2 \leq 1$ and

$$f'(t) = 3t^2 - 2t - y^2 \geq 3t^2 - 2t - 1 = (3t + 1)(t - 1) \geq 0$$

for all real values $t \geq 1$. This implies that the derivative $f'(t)$ is positive on the interval $[1, \infty)$ and, therefore, the function $f(t)$ increases on $[1, \infty)$. This in turn implies that $t = 1$ is the minimum of the function $f(t)$ on $[1, \infty)$. Therefore

$$
\begin{aligned}
f(t) &\geq f(1) \\
&= (1)^3 - (1)^2 - y^2(1) + y^3 - y + 1 \\
&= y^3 - y^2 - y + 1 \\
&= (y - 1)^2(y + 1)
\end{aligned}
$$

Notice that since $y \in (0, 1]$, then the last expression equals zero only for $y = 1$ and is nonnegative for all other y. Therefore, we have that $f(t) \geq 0$ or equivalently

$$t^3 - t^2 - y^2 t + y^3 - y + 1 \geq 0$$

Substituting t for x we have the inequality

$$x^3 - x^2 - y^2 x + y^3 - y + 1 \geq 0$$

where the equality holds only when $y = 1$ and $x = 1$. From here we have that $a = b = c$.

It is not hard to check that the triple of equal numbers indeed satisfies the initial equation.[2]

---

[2]The same conclusion can be obtained by applying the Rearrangement Inequality discussed in detail in Chapter 30.

Problem 3

Given the positive real numbers $a \geq b \geq c$, such that

$$(a + b + c)^3 = 64abc$$

Show that

$$2(ab + bc + ac) \geq a^2 + b^2 - 3c^2$$

Solution

Since $c \neq 0$, let us divide both sides of the inequality by c^2:

$$\frac{2(ab + bc + ac)}{c^2} = \frac{a^2 + b^2 - 3c^2}{c^2}$$

$$2\left(\frac{ab}{c^2} + \frac{bc}{c^2} + \frac{ac}{c^2}\right) = \frac{a^2}{c^2} + \frac{b^2}{c^2} - \frac{3c^2}{c^2}$$

$$2\left(\frac{a}{c} \cdot \frac{b}{c} + \frac{b}{c} + \frac{a}{c}\right) = \left(\frac{a}{c}\right)^2 + \left(\frac{b}{c}\right)^2 - 3$$

Let us make the following substitutions:

$$\frac{a}{c} = x$$

$$\frac{b}{c} = y$$

Therefore, the inequality can be rewritten as follows:

$$2(xy + x + y) \geq x^2 + y^2 - 3$$

or equivalently

$$4xy + 2(x + y) + 3 \geq (x + y)^2$$

Now let us divide both sides of the given equation by c^3:

$$(a + b + c)^3 = 64abc$$

$$\frac{(a + b + c)^3}{c^3} = \frac{64abc}{c^3}$$

$$\left(\frac{a}{c} + \frac{b}{c} + 1\right)^3 = 64 \cdot \frac{a}{c} \cdot \frac{b}{c}$$

$$(x + y + 1)^3 = 64xy$$

Let us now put

$$x + y = M$$

$$xy = N$$

The problem can now be rewritten as follows: prove the inequality

$$4N + 2M + 3 \geq M^2$$

given that

$$(M + 1)^3 = 64N$$

Multiplying the needed inequality by 16 we have an equivalent inequality:

$$64N + 32M + 48 \geq 16M^2$$

Now substituting $64N = (M + 1)^3$ we have

$$(M + 1)^3 + 32M + 48 \geq 16M^2$$
$$M^3 + 3M^2 + 35M + 49 \geq 16M^2$$
$$M^3 - 13M^2 + 35M + 49 \geq 0$$

Let us now consider the function $f(t) = t^3 - 13t^2 + 35t + 49$ on the interval $[-1, \infty)$. Its first derivative is

$$f'(t) = \left(t^3 - 13t^2 + 35t + 49\right)' = 3t^2 - 26t + 35$$

and its critical points are $t = \frac{5}{3}$ and $t = 7$.

The derivative $f'(t)$ is positive on the interval $\left[-1, \frac{5}{3}\right)$ and, therefore, the function $f(t)$ increases on $\left[-1, \frac{5}{3}\right]$. The derivative $f'(t)$ is negative on the interval $\left(\frac{5}{3}, 7\right)$ and, therefore, the function $f(t)$ decreases on $\left[\frac{5}{3}, 7\right]$. The derivative $f'(t)$ is positive on the interval $(7, \infty)$ and, therefore, the function $f(t)$ increases on $[7, \infty)$.

This implies that $f(t) \geq f(-1)$ for $t \in \left[-1, \frac{5}{3}\right]$ and $f(t) \geq f(7)$ for $t \in \left[\frac{5}{3}, \infty\right)$. However, since

$$f(-1) = f(7) = 0$$

then

$$f(t) \geq 0$$

which is what needed to be proven.

CHAPTER 35

Increasing and Decreasing Functions

Numerous math olympiad problems rely on demonstrating that a particular function is increasing or decreasing over specific interval. Under certain conditions these functions take all their values exactly once and prove highly beneficial in establishing the uniqueness of solutions of equations.

Increasing and Decreasing Functions

Function $f(x)$ is called **increasing** on the interval (a, b) if for all x_1 and x_2 from (a, b), such that $x_1 < x_2$ it holds that $f(x_1) < f(x_2)$.

Function $f(x)$ is called **decreasing** on the interval (a, b) if for all x_1 and x_2 from (a, b), such that $x_1 < x_2$ it holds that $f(x_1) > f(x_2)$.

The increasing and decreasing properties of the linear, quadratic, exponential, logarithmic and trigonometric functions are very well-known. For other functions we recommend to use the first derivative.

First Derivative Test

The intervals where the function $f(x)$ is increasing or decreasing are determined from the following:

- If $f'(x)$ is positive on the interval (a, b), then $f(x)$ is increasing on the interval (a, b).

- If $f'(x)$ is negative on the interval (a, b), then $f(x)$ is decreasing on the interval (a, b).

Let us consider several problems.

Problem 1

Solve the equation in real numbers

$$5^x + 12^x = 13^x$$

Solution

Notice that $x = 2$ satisfies the equation, since

$$5^2 + 12^2 = 13^2$$

Let us show that it is the only solution.

Since $13^x \neq 0$, let us divide both sides of the equation by 13^x and rewrite it as follows:

$$\frac{5^x}{13^x} + \frac{12^x}{13^x} = \frac{13^x}{13^x}$$

$$\left(\frac{5}{13}\right)^x + \left(\frac{12}{13}\right)^x = 1$$

Let us now consider the exponential functions $f(x) = \left(\frac{5}{13}\right)^x$ and $g(x) = \left(\frac{12}{13}\right)^x$.

Notice that since their bases are less than 1, then both $f(x)$ and $g(x)$ are strictly decreasing functions, and so is their sum. This implies that the left-hand side of the last equation represents a strictly decreasing function, which takes each of its values exactly once. Therefore, it takes the value 1 only at $x = 2$, which implies that $x = 2$ is the only solution of the initial equation.

Problem 2

Given positive real numbers a, b and c. Find all real numbers x, such that

$$\sqrt[3]{\frac{a}{x}+b}+\sqrt[3]{\frac{b}{x}+c}+\sqrt[3]{\frac{c}{x}+a}=\sqrt[3]{ax+b}+\sqrt[3]{bx+c}+\sqrt[3]{cx+a}$$

Solution

Notice that for $x=1$ the equation becomes an identity

$$\sqrt[3]{a+b}+\sqrt[3]{b+c}+\sqrt[3]{c+a}=\sqrt[3]{a+b}+\sqrt[3]{b+c}+\sqrt[3]{c+a}$$

and, therefore, $x=1$ is a solution. We will show that it is the only solution of the initial equation. Let us rewrite the equation as follows:

$$\sqrt[3]{\frac{a}{x}+b}+\sqrt[3]{\frac{b}{x}+c}+\sqrt[3]{\frac{c}{x}+a}-\sqrt[3]{ax+b}-\sqrt[3]{bx+c}-\sqrt[3]{cx+a}=0$$

Let us now consider the functions

$$f(x)=\sqrt[3]{\frac{a}{x}+b}+\sqrt[3]{\frac{b}{x}+c}+\sqrt[3]{\frac{c}{x}+a}$$

$$g(x)=-\sqrt[3]{ax+b}-\sqrt[3]{bx+c}-\sqrt[3]{cx+a}$$

Notice that both $f(x)$ and $g(x)$ are strictly decreasing functions, and so is their sum. This implies that the left-hand side of the last equation represents a strictly decreasing function, which takes each of its values exactly once. Therefore, it takes the value 0 only at $x=1$, which implies that $x=1$ is the only solution of the initial equation.

Problem 3

Solve the equation

$$e^{\left(x^{\frac{1}{c}}\right)}\cdot x^{\left(\frac{1}{c}\right)^x}=1$$

in positive numbers x.

Solution

Notice that $x = \frac{1}{e}$ satisfies the equation:

$$e^{\left(\frac{1}{e}\right)^{\frac{1}{e}}} \cdot \left(\frac{1}{e}\right)^{\left(\frac{1}{e}\right)^{\frac{1}{e}}} = e^{\left(\frac{1}{e}\right)^{\frac{1}{e}}} \cdot e^{-\left(\frac{1}{e}\right)^{\frac{1}{e}}}$$

$$= e^{\left(\frac{1}{e}\right)^{\frac{1}{e}} - \left(\frac{1}{e}\right)^{\frac{1}{e}}}$$

$$= e^{0}$$

$$= 1$$

Let us show that it is the only solution.

Let us take the natural logarithm of both sides of the equation and rewrite it as

$$\ln\left(e^{\left(x^{\frac{1}{e}}\right)} \cdot x^{\left(\frac{1}{e}\right)^{x}}\right) = \ln(1)$$

$$\ln\left(e^{\left(x^{\frac{1}{e}}\right)}\right) + \ln\left(x^{\left(\frac{1}{e}\right)^{x}}\right) = 0$$

$$x^{\frac{1}{e}} + \left(\frac{1}{e}\right)^{x}\ln x = 0$$

$$x^{\frac{1}{e}}e^{x} + \ln x = 0$$

Let us now consider the function

$$f(x) = x^{\frac{1}{e}}e^{x} + \ln x$$

Its first derivative is

$$f'(x) = \left(x^{\frac{1}{e}}e^{x} + \ln x\right)' = \frac{e^{x}}{x^{\frac{e-1}{e}}} + x^{\frac{1}{e}}e^{x} + \frac{1}{x}$$

Notice that $f'(x) > 0$ for all positive x. This implies that the left-hand side of the last equation represents a strictly increasing function, which takes each of its values exactly once. Therefore, it takes the value 0 only at $x = \frac{1}{e}$, which implies that $x = \frac{1}{e}$ is the only solution of the initial equation.

CHAPTER 36

Maximum and Minimum of a Function

Many math olympiad problems rely on demonstrating that a particular function has a maximum or a minimum on a specific interval. This technique is highly useful in proving inequalities and establishing the uniqueness of solutions of equations.

Maximum and Minimum of a Function

Number $x = c$ is called the **maximum** of the function $f(x)$ on the interval (a, b) if for all x from (a, b), it holds that $f(x) \leq f(c)$.

Number $x = c$ is called the **minimum** of the function $f(x)$ on the interval (a, b) if for all x from (a, b), it holds that $f(x) \geq f(c)$.

The maximum and minimum properties of the linear, quadratic, exponential, logarithmic and trigonometric functions are very well-known. For other functions we recommend to use the first derivative.

Critical Point

Number $x = c$ is called the **critical point** of the function $f(x)$ if the value of the function $f(c)$ exists and the value of the derivative $f'(c)$ is zero or does not exist.

First Derivative Test

Let $x = c$ be the only **critical point** of the function $f(x)$ on the interval (a, b).

- If $f'(x)$ changes its sign from "+" to "−" at $x = c$, then $x = c$ is the maximum of the function $f(x)$ on the interval (a, b).

- If $f'(x)$ changes its sign from "−" to "+" at $x = c$, then $x = c$ is the minimum of the function $f(x)$ on the interval (a, b).

Let us consider several problems.

Problem 1

Which is larger e^π or π^e?

Solution

Let us show that

$$e^\pi > \pi^e$$

Let us take the natural logarithm of both sides of the inequality and rewrite it as

$$\ln(e^\pi) > \ln(\pi^e)$$
$$\pi > e \ln \pi$$
$$\pi - e \ln \pi > 0$$

Let now us consider the function $f(x) = x - e \ln x$ on the interval $(0, \infty)$. It's first derivative is

$$f'(x) = (x - e \ln x)' = 1 - \frac{e}{x}$$

and it's only critical point is $x = e$. Since $f'(x) < 0$ for all $x \in (0, e)$, then $f(x)$ is decreasing on $(0, e)$. Since $f'(x) > 0$ for all $x \in (e, \infty)$, then $f(x)$ is increasing on (e, ∞). This implies that $x = e$ is the minimum of the function $f(x)$ on the interval $(0, \infty)$.

Therefore, $f(\pi) > f(e)$ and

$$\pi - e \ln \pi = f(\pi) > f(e) = e - e \ln e = 0$$

as desired.

Problem 2

Given the positive integers n and m, such that $n > m$. Find all positive x, such that

$$\frac{x^n + x^{n-1} + \ldots + x + 1}{x^m + x^{m-1} + \ldots + x + 1} = \frac{n+1}{m+1}$$

Solution

Notice that $x = 1$ satisfies the equation. From now on we will assume that $x \neq 1$.

Let us multiply the numerator and the denominator of the left-hand side by $(x - 1)$ and rewrite the inequality as follows:

$$\frac{x^n + x^{n-1} + \ldots + x + 1}{x^m + x^{m-1} + \ldots + x + 1} = \frac{n+1}{m+1}$$

$$\frac{(x-1)\left(x^n + x^{n-1} + \ldots + x + 1\right)}{(x-1)\left(x^m + x^{m-1} + \ldots + x + 1\right)} = \frac{n+1}{m+1}$$

$$\frac{x^{n+1} - 1}{x^{m+1} - 1} = \frac{n+1}{m+1}$$

$$\left(x^{n+1} - 1\right)(m+1) = \left(x^{m+1} - 1\right)(n+1)$$

$$(m+1)x^{n+1} - m - 1 = (n+1)x^{m+1} - n - 1$$

$$(m+1)x^{n+1} - (n+1)x^{m+1} = m - n$$

We will prove that for positive $x \neq 1$ the following inequality holds:

$$(m+1)x^{n+1} - (n+1)x^{m+1} > m - n$$

Let us consider the function $f(x) = (m+1)x^{n+1} - (n+1)x^{m+1}$ on the interval $(0, \infty)$. It's first derivative is

$$f'(x) = \left((m+1)x^{n+1} - (n+1)x^{m+1}\right)'$$

$$= (m+1)(n+1)x^n - (m+1)(n+1)x^m$$

$$= (m+1)(n+1)x^m \left(x^{n-m} - 1\right)$$

The only critical point of the function $f(x)$ on the interval $(0, \infty)$ is $x = 1$. Since $f'(x) < 0$ for all $x \in (0, 1)$, then $f(x)$ is decreasing on $(0, 1)$. Since $f'(x) > 0$

all $x \in (1, \infty)$, then $f(x)$ is increasing on $(1, \infty)$. This implies that $x = 1$ is the minimum of the function $f(x)$ on the interval $(0, \infty)$. Therefore, $f(x) > f(1)$ for $x \neq 1$ and

$$(m+1) x^{n+1} - (n+1) x^{m+1} = f(x) > f(1) = (m+1) - (n+1) = m - n$$

as desired.

We conclude that $x = 1$ is the only solution.

Problem 3

Find all positive real numbers x that satisfy the equation

$$4^x - 5x = 6 - 7^x$$

Solution

Let us rewrite the equation as follows

$$4^x + 7^x - 5x - 6 = 0$$

Notice that $x = 1$ satisfies the equation, since

$$4^1 + 7^1 - 5(1) - 6 = 0$$

Let us prove that $x = 1$ is the only solution. Let us consider the function

$$f(x) = 4^x + 7^x - 5x - 6$$

on the interval $(0, \infty)$. It's first derivative is

$$f'(x) = (4^x + 7^x - 5x - 6)' = 4^x \ln 4 + 7^x \ln 7 - 5$$

Let us show that the $f(x)$ has a unique critical point x_0 on the interval $(0, \infty)$.

First, we will prove that $f'(0) < 0$ and $f'(1) > 0$. Indeed, we have

$$f'(0) = \ln 4 + \ln 7 - 5 = \ln 28 - 5 < \ln 32 - 5 = 5 \ln 2 - 5 < 5 \ln e - 5 = 0$$

and

$$f'(1) = 4 \ln 4 + 7 \ln 7 - 5 > 7 \ln 7 - 5 > 5 \ln 7 - 5 > 5 \ln e - 5 = 0$$

Therefore, by the Intermediate Value Theorem $f'(x)$ has a real root on the interval $(0, 1)$.

Now let us show that this root is unique. The second derivative of $f(x)$ is equal to

$$f''(x) = (4^x \ln 4 + 7^x \ln 7 - 5)' = 4^x \ln 16 + 7^x \ln 49$$

and is positive for all real values of x. This implies that $f'(x)$ is strictly increasing and takes each of its values exactly once. From here we conclude that the real root of $f'(x)$ is unique, and, therefore, $f(x)$ has a unique critical point on the interval $(0, \infty)$. Furthermore, since $f''(x) > 0$, then the critical point of $f(x)$ is the minimum.

From here we have that if $x \in [0, x_0]$, then

$$f(x) \leq f(0) = -4 < 0$$

and, therefore, there are no solutions on $[0, x_0]$.

However, if $x \in (x_0, \infty)$, then $f'(x)$ is positive and $f(x)$ takes each of its values exactly once, which implies that $x = 1$ is the only solution.

CHAPTER 37

Mathematical Induction

Mathematical Induction is a go-to method for proving that a statement is true for all natural numbers n.

Typical Steps of Mathematical Induction

- **Basis of Induction**. Prove that the statement of the problem is true for some initial value, for example $n = 1$.

- **Inductive Step**. Assume that the statement is true for some $n = k$, and then prove that it is also true for $n = k + 1$.

By combining the **Basis of Induction** and the **Inductive Step**, we can conclude that the statement is true for all positive integers n.

In some problems the initial value can be chosen to be larger than 1. Sometimes a stronger form of induction might be necessary, where the inductive step depends on several values of k, or even on all the previous values together.

Let us consider several problems.

Problem 1

Let x be a real number, such that $x + x^{-1}$ is an integer. Show that

$$x^n + x^{-n}$$

is also an integer for all $n \in \mathbb{N}$.

Solution

Even though this problem involves integer numbers, its solution is very algebraic. Let us prove the following claim by induction on n.

Claim

If $x + x^{-1}$ is an integer, then $x^n + x^{-n}$ is also an integer for all $n \in \mathbb{N}$.

Proof

- **Basis of Induction**. We will prove that the claim is true for $n = 1$ and $n = 2$.

  For $n = 1$ the claim obviously holds, since $x + x^{-1}$ is an integer.

  For $n = 2$ we have

  $$x^2 + x^{-2} = \left(x + x^{-1}\right)^2 - 2$$

  and, therefore, is also an integer.

- **Inductive Step**. Let us assume that the claim holds for $n = k - 1$ and $n = k$, i.e. that $x^{k-1} + x^{-(k-1)}$ and $x^k + x^{-k}$ are integer numbers.

  Let us consider the expression

  $$\left(x^k + x^{-k}\right)\left(x + x^{-1}\right) = x^{k+1} + x^{-(k+1)} + x^{k-1} + x^{-(k-1)}$$

  From here

  $$x^{k+1} + x^{-(k+1)} = \left(x^k + x^{-k}\right)\left(x + x^{-1}\right) - x^{k-1} - x^{-(k-1)}$$

  is an integer number and the claim holds for $n = k + 1$. ∎

We conclude that the statement of the problem is true for all positive integers n.

Problem 2

Show that for all $n \in \mathbb{N}$

$$1 \cdot 3 \cdot 5 \cdot \ldots \cdot (2n - 1) \leq n^n$$

Solution

Let us prove the following claim by induction on n.

Claim

Show that for all $n \in \mathbb{N}$

$$1 \cdot 3 \cdot 5 \cdot \ldots \cdot (2n - 1) \leq n^n$$

Proof

- **Basis of Induction**. The claim holds for $n = 1$ since

$$1 \leq 1^1$$

- **Inductive Step**. Let us assume that the claim holds for $n = k$:

$$1 \cdot 3 \cdot 5 \cdot \ldots \cdot (2k - 1) \leq k^k$$

Start by noticing that for all positive integers k

$$\frac{2k + 1}{k + 1} < \frac{2k + 2}{k + 1} = \frac{2(k + 1)}{k + 1} = 2$$

Furthermore, by Bernoulli's Inequality:

$$\left(1 + \frac{1}{k}\right)^k \geq 1 + \frac{1}{k} \cdot k = 2$$

From the last two inequalities we have

$$\frac{2k+1}{k+1} < \left(1 + \frac{1}{k}\right)^k$$

$$\frac{2k+1}{k+1} < \left(\frac{k+1}{k}\right)^k$$

$$\frac{2k+1}{k+1} < \frac{(k+1)^k}{k^k}$$

$$2k+1 < \frac{(k+1)^{k+1}}{k^k}$$

Now let us apply the inductive hypothesis to the expression

$$1 \cdot 3 \cdot 5 \cdot \ldots \cdot (2k-1) \cdot (2k+1) \leq k^k \cdot (2k+1)$$

$$= k^k \cdot \frac{(k+1)^{k+1}}{k^k}$$

$$= (k+1)^{k+1}$$

and the statement holds for $n = k + 1$.   ∎

We conclude that the statement of the problem is true for all positive integers n.

Problem 3

The sequences a_n and b_n are defined as $a_1 = 1, b_1 = 2$ and

$$a_{n+1} = \frac{1 + a_n + a_n b_n}{b_n}$$

$$b_{n+1} = \frac{1 + b_n + a_n b_n}{a_n}$$

Prove that $a_n < 5$ for all positive integers n.

Solution

Let us prove the following claim by induction on n.

Claim

For all positive integers n:

$$\frac{1}{a_n + 1} - \frac{1}{b_n + 1} = \frac{1}{6}$$

Proof

- **Basis of Induction**. The claim holds for $n = 1$ since

$$\frac{1}{1+1} - \frac{1}{2+1} = \frac{1}{2} - \frac{1}{3} = \frac{1}{6}$$

- **Inductive Step**. Let us assume that the claim holds for $n = k$:

$$\frac{1}{a_k + 1} - \frac{1}{b_k + 1} = \frac{1}{6}$$

Let us consider the expression

$$
\begin{aligned}
\frac{1}{a_{k+1} + 1} - \frac{1}{b_{k+1} + 1} &= \frac{1}{\frac{1 + a_k + a_k b_k}{b_k} + 1} - \frac{1}{\frac{1 + b_k + a_k b_k}{a_k} + 1} \\
&= \frac{b_k}{1 + a_k + b_k + a_k b_k} - \frac{a_k}{1 + a_k + b_k + a_k b_k} \\
&= \frac{b_k - a_k}{(1 + a_k)(1 + b_k)} \\
&= \frac{(b_k + 1) - (a_k + 1)}{(a_k + 1)(b_k + 1)} \\
&= \frac{1}{a_k + 1} - \frac{1}{b_k + 1} \\
&= \frac{1}{6}
\end{aligned}
$$

and the claim holds for $n = k + 1$. ∎

Since

$$\frac{1}{a_n + 1} - \frac{1}{b_n + 1} = \frac{1}{6}$$

then

$$\frac{1}{a_n + 1} = \frac{1}{b_n + 1} + \frac{1}{6} > \frac{1}{6}$$

This implies that

$$a_n + 1 < 6$$

or equivalently $a_n < 5$, as desired.

CHAPTER 38

Functional Equations and Simple Substitutions

In this chapter we will focus on the functional equations that are solved by several consequent **simple substitutions**. The obtained equations are then combined to find the explicit formula of the unknown function.

We recommend to start by trying the following substitutions: $x \to 0, y \to 0, x \to \pm 1$, $y \to \pm 1, y \to x, y \to -x, x \to -x, y \to -y$. Other substitutions might also be necessary depending on the structure of the problem.

It is important to highlight that once we find a function that is a solution, we need to substitute it into the original functional equation and verify if it indeed satisfies the conditions of the problem.

Let us consider several problems.

Problem 1

Find all functions $f : \mathbb{R} \to \mathbb{R}$, such that for all $x, y \in \mathbb{R}$

$$f\left(x^2 + y\right) - xf(x + y) = 0$$

Solution

Let us start by substituting $x \to 0$ and $y \to 0$ into the original equation:

$$f\left((0)^2 + 0\right) + (0)f\left((0) + 0\right) = 0$$
$$f(0) = 0$$

Now let us put $x \neq 0$ and substitute $x \to -x$ into the original equation:

$$f\left((-x)^2 + y\right) + xf\left((-x) + y\right) = 0$$
$$f\left(x^2 + y\right) + xf(y - x) = 0$$

Subtracting this equation from the original equation we have

$$f\left(x^2 + y\right) - xf(x + y) - f\left(x^2 + y\right) - xf(y - x) = 0$$
$$-xf(x + y) - xf(y - x) = 0$$
$$-x\left(f(x + y) - f(y - x)\right) = 0$$
$$f(x + y) - f(y - x) = 0$$
$$f(x + y) = f(y - x)$$

Substituting $y \to x$ into the last equation we have:

$$f(2x) = f(x - x) = f(0) = 0$$

Now substituting $2x \to t$ and taking into account that $f(0) = 0$ we have that for all real values of t

$$f(t) = 0$$

It is not hard to check that this function indeed satisfies the original equation.

Problem 2

Find all functions $f : \mathbb{R} \to \mathbb{R}$, such that for all $x, y \in \mathbb{R}$

$$f(x + y) - 2f(x - y) + f(x) - 2f(y) = y - 2$$

Solution

Let us start by substituting $y \to 0$ into the original equation:

$$f\left(x + (0)\right) - 2f\left(x - (0)\right) + f(x) - 2f(0) = (0) - 2$$
$$f(x) - 2f(x) + f(x) - 2f(0) = -2$$
$$-2f(0) = -2$$
$$f(0) = 1$$

Now let us substitute $x \to 0$ into the original equation:

$$f((0) + y) - 2f((0) - y) + f(0) - 2f(y) = y - 2$$
$$f(y) - 2f(-y) + 1 - 2f(y) = y - 2$$
$$-f(y) - 2f(-y) = y - 3$$
$$f(y) + 2f(-y) = 3 - y$$

Substituting $y \to -y$ into the last equation we have

$$f(-y) + 2f(y) = 3 + y$$

Together with the previous equation we have the following system of two equations:

$$\begin{cases} f(y) + 2f(-y) = 3 - y \\ f(-y) + 2f(y) = 3 + y \end{cases}$$

Multiplying the second equation by -2 and adding with the first equation we have

$$f(y) + 2f(-y) - 2f(-y) - 4f(y) = 3 - y - 6 - 2y$$
$$-3f(y) = -3 - 3y$$
$$f(y) = y + 1$$

It is not hard to check that this function indeed satisfies the original equation.

Problem 3

Find all functions $f : \mathbb{R} \to \mathbb{R}$, such that for all $x, y \in \mathbb{R}$

$$f\left(x + y^2\right) = f\left(x^2\right) + f(y)$$

Solution

Let us start by substituting $x \to 0$ and $y \to 0$ into the original equation:

$$f\left((0) + (0)^2\right) = f\left((0)^2\right) + f(0)$$
$$f(0) = f(0) + f(0)$$
$$f(0) = 0$$

Now let us substitute only $y \to 0$ into the original equation:

$$f\left(x + (0)^2\right) = f\left(x^2\right) + f(0)$$
$$f(x) = f\left(x^2\right)$$

Substituting $x \to -x$ into the last equation we have

$$f(-x) = f\left((-x)^2\right) = f\left(x^2\right) = f(x)$$

and the function $f(x)$ is even.

Let us substitute $x \to y^2$ into the original equation:

$$f\left((y^2) + y^2\right) = f\left((y^2)^2\right) + f(y)$$
$$f\left(2y^2\right) = f\left(y^4\right) + f(y)$$

Now we will substitute $x \to -y^2$ into the original equation:

$$f\left((-y^2) + y^2\right) = f\left((-y^2)^2\right) + f(y)$$
$$f(0) = f\left(y^4\right) + f(y)$$
$$0 = f\left(y^4\right) + f(y)$$
$$0 = f\left(2y^2\right)$$

Notice that the last equality is true for all real values of y. However, when y runs through the real numbers the expression $2y^2$ covers all numbers from the interval $[0, \infty)$. Therefore, for all nonnegative t we have $f(t) = 0$. Since the function $f(t)$ is also even, then for all negative t we have

$$f(t) = f(-t) = 0$$

and, therefore, $f(t) = 0$ for all real numbers.

It is not hard to check that this function indeed satisfies the original equation.

CHAPTER 39

Functional Equations and Cyclic Substitutions

In this chapter we will focus on the functional equations that are solved using the **cyclic substitutions**. These are the series of substitutions of the variable x by the same expression. This usually results in a cyclic set of arguments for the original function and produces a system of two or more equations.

It is important to highlight that once we find a function that is a solution, we need to substitute it into the original function equation and verify if it indeed satisfies the conditions of the problem.

Let us consider several problems.

Problem 1

Find all functions $f : \mathbb{R} \backslash \{0\} \to \mathbb{R}$, such that for all $x \neq 0$

$$xf(x) + 2f\left(-\frac{1}{x}\right) = 3$$

Solution

Let us substitute x for $-\frac{1}{x}$ into the original equation:

$$-\frac{1}{x}f\left(-\frac{1}{x}\right) + 2f(x) = 3$$

Multiplying both sides of this equation by x we have

$$-f\left(-\frac{1}{x}\right) + 2xf(x) = 3x$$

Together with the first equation we have the following system of two equations:

$$\begin{cases} xf(x) + 2f\left(-\frac{1}{x}\right) = 3 \\ -f\left(-\frac{1}{x}\right) + 2xf(x) = 3x \end{cases}$$

Multiplying the second equation by 2 and adding with the first equation we have

$$xf(x) + 2f\left(-\frac{1}{x}\right) - 2f\left(-\frac{1}{x}\right) + 4xf(x) = 3 + 6x$$

$$5xf(x) = 3 + 6x$$

$$f(x) = \frac{3 + 6x}{5x}$$

It is not hard to check that this function indeed satisfies the original equation.

Problem 2

Find the function $f : \mathbb{R}\backslash\{1\} \to \mathbb{R}$, that satisfies the following functional equation for all $x \neq 1$:

$$(x - 1)f\left(\frac{x+1}{x-1}\right) = ax + f(x)$$

where a is a real constant.

Solution

Let us substitute x for $\frac{x+1}{x-1}$ into the original equation:

$$\left(\frac{x+1}{x-1} - 1\right)f(x) = a\left(\frac{x+1}{x-1}\right) + f\left(\frac{x+1}{x-1}\right)$$

Multiplying both sides of this equation by $(x - 1)$ we have

$$2f(x) = a(x + 1) + (x - 1)f\left(\frac{x + 1}{x - 1}\right)$$

Together with the first equation we have the following system of two equations:

$$\begin{cases} (x - 1)f\left(\dfrac{x + 1}{x - 1}\right) = x + f(x) \\ 2f(x) = a(x + 1) + (x - 1)f\left(\dfrac{x + 1}{x - 1}\right) \end{cases}$$

Adding these equations we have

$$(x - 1)f\left(\frac{x + 1}{x - 1}\right) + 2f(x) = x + f(x) + a(x + 1) + (x - 1)f\left(\frac{x + 1}{x - 1}\right)$$

$$2f(x) = x + f(x) + a(x + 1)$$

$$f(x) = (a + 1)x + a$$

It is not hard to check that this function indeed satisfies the original equation.

Problem 3

Find all functions $f : \mathbb{R} \to \mathbb{R}$, such that for all $x \neq 0, x \neq 1$

$$f(x) + f\left(\frac{1}{1 - x}\right) = x$$

Solution

Let us substitute x for $\frac{1}{1-x}$ into the original equation:

$$f\left(\frac{1}{1 - x}\right) + f\left(\frac{x - 1}{x}\right) = \frac{1}{1 - x}$$

Let us substitute x for $\frac{1}{1-x}$ into the last equation:

$$f\left(\frac{x - 1}{x}\right) + f(x) = \frac{x - 1}{x}$$

Let us introduce the following variables:

$$f(x) = a$$

$$f\left(\frac{1}{1 - x}\right) = b$$

$$f\left(\frac{x - 1}{x}\right) = c$$

Then we have the following system of equations:

$$\begin{cases} a + b = x \\ b + c = \dfrac{1}{1-x} \\ c + a = \dfrac{x-1}{x} \end{cases}$$

We are interested in finding the expression for a. Adding all three equations we have

$$2(a + b + c) = x + \frac{1}{1-x} + \frac{x-1}{x}$$

$$a + b + c = \frac{1}{2}\left(x + \frac{1}{1-x} + \frac{x-1}{x} \right)$$

Now we can find a by substituting $b + c = \frac{1}{1-x}$ into the last equation:

$$a + \left(\frac{1}{1-x} \right) = \frac{1}{2}\left(x + \frac{1}{1-x} + \frac{x-1}{x} \right)$$

$$a = \frac{1}{2}\left(x - \frac{1}{1-x} + \frac{x-1}{x} \right)$$

$$a = \frac{x^3 - x + 1}{2x^2 - 2x}$$

and we have that

$$f(x) = \frac{x^3 - x + 1}{2x^2 - 2x}$$

It is not hard to check that this function indeed satisfies the original equation.

CHAPTER 40

Functional Equations on Integer Domain

In this chapter we will focus on the functional equations on natural or integer domains. Induction is very useful for dealing with positive integer arguments. The case of zero is usually dealt apart. For negative integer arguments we need to establish either the even/odd properties of the function or some similar identities.

It is important to highlight that once we find a function that is a solution, we need to substitute it into the original function equation and verify if it indeed satisfies the conditions of the problem.

Let us consider several problems.

Problem 1

Prove that there are no functions $f : \mathbb{N} \to \mathbb{N}$, such that $f(1) = 1$ and

$$f(f(n)) + f^2(n + m) = n^2 + 3n + m^2$$

for all positive integers m and n.

Solution

Let us start by substituting $m \to 1$ into the original equation:

$$f(f(n)) + f^2(n+1) = n^2 + 3n + 1$$

Let us prove the following claim by induction on n.

Claim

Let $f : \mathbb{N} \to \mathbb{N}$, such that $f(1) = 1$. Given that for all $n \in \mathbb{N}$:

$$f(f(n)) + f^2(n+1) = n^2 + 3n + 1$$

then $f(n) = n$ for all $n \in \mathbb{N}$.

Proof

- **Basis of Induction.** The claim is true for $n = 1$, since $f(1) = 1$.

- **Inductive Step.** Let us assume that the claim holds for $n = k$, i.e. that $f(k) = k$. Let us substitute $n \to k$ into the functional equation:

$$f(f(k)) + f^2(k+1) = k^2 + 3k + 1$$
$$f(k) + f^2(k+1) = k^2 + 3k + 1$$
$$k + f^2(k+1) = k^2 + 3k + 1$$
$$f^2(k+1) = k^2 + 2k + 1$$
$$f^2(k+1) = (k+1)^2$$

Since the codomain of the function $f(n)$ is $\mathbb{N}$, then $f(k+1) = k+1$ and the claim holds for $n = k+1$. ∎

Now we will verify if $f(n) = n$ satisfies the original equation:

$$f(f(n)) + f^2(n+m) = n^2 + 3n + m^2$$
$$n + (n+m)^2 = n^2 + 3n + m^2$$
$$n + n^2 + mn + m^2 = n^2 + 3n + m^2$$
$$n(m-2) = 0$$

which does not hold for $m \neq 2$.

We conclude that there are no functions that satisfy the conditions of the problem.

Problem 2

Find all functions $f : \mathbb{N} \to \mathbb{N}$ satisfying

$$f\left(f(n)\right) + f(n) = 2n + 3$$

Solution

Answer: $f(n) = n + 1$.

Let us start by substituting $n \to 1$ into the original equation:

$$f\left(f(1)\right) + f(1) = 2(1) + 3$$
$$f\left(f(1)\right) + f(1) = 5$$

Notice that both numbers $f\left(f(1)\right)$ and $f(1)$ are positive integers and, therefore, $f(1)$ may only take values 1, 2, 3 or 4.

We will proceed by doing the following casework:

- If $f(1) = 1$, then substituting $n \to 1$ into the original equation we have

$$f\left(f(1)\right) + f(1) = 2(1) + 3$$
$$f(1) + 1 = 5$$
$$1 + 1 = 5$$

  which leads to a contradiction.

- If $f(1) = 2$, then let us prove the following claim by induction on n.

  ### Claim

  Let $f : \mathbb{N} \to \mathbb{N}$, such that $f(1) = 2$. Given that for all $n \in \mathbb{N}$:

  $$f\left(f(n)\right) + f(n) = 2n + 3$$

  then $f(n) = n + 1$ for all $n \in \mathbb{N}$.

  ### Proof

  - **Basis of Induction.** The claim is true for $n = 1$, since $f(1) = 2$.

  - **Inductive Step.** Let us assume that the claim holds for $n = k$, i.e. that $f(k) = k + 1$.

Let us substitute $n \to k$ into the functional equation:

$$f\left(f(k)\right) + f(k) = 2k + 3$$
$$f\left(k + 1\right) + k + 1 = 2k + 3$$
$$f\left(k + 1\right) = k + 2$$

and the claim holds for $n = k + 1$. ∎

It is not hard to check that the function $f(n) = n + 1$ satisfies the original equation.

- If $f(1) = 3$, then substituting $n \to 1$ into the original equation we have

$$f\left(f(1)\right) + f(1) = 2(1) + 3$$
$$f(3) + 3 = 5$$
$$f(3) = 2$$

Now substituting $n \to 3$ into the original equation we have

$$f\left(f(3)\right) + f(3) = 2(3) + 3$$
$$f(2) + 2 = 9$$
$$f(2) = 7$$

And finally substituting $n \to 2$ into the original equation we have

$$f\left(f(2)\right) + f(2) = 2(2) + 3$$
$$f(7) + 7 = 7$$
$$f(7) = 0$$

which leads to a contradiction.

- If $f(1) = 4$, then substituting $n \to 1$ into the original equation we have

$$f\left(f(1)\right) + f(1) = 2(1) + 3$$
$$f(4) + 4 = 5$$
$$f(4) = 1$$

Now substituting $n \to 4$ into the original equation we have

$$f\left(f(4)\right) + f(4) = 2(4) + 3$$
$$f(1) + 1 = 11$$
$$4 + 1 = 11$$

which leads to a contradiction.

Problem 3

Find all functions $f : \mathbb{Z} \to \mathbb{Z}$, such that for all $m, n \in \mathbb{Z}$

$$f(m)f(n) = f(m + n)$$

Solution

Let us start by substituting $m \to 0$ and $n \to 0$ into the original equation:

$$f(0)f(0) = f(0 + 0)$$
$$f(0)f(0) = f(0)$$
$$f(0)\left(f(0) - 1\right) = 0$$

We will proceed by doing the following casework:

- If $f(0) = 0$, then we will substitute $m \to 0$ into the original equation:

$$f(0)f(n) = f(0 + n)$$

from where $f(n) = 0$ for all integers n. It is not hard to check that this function satisfies the original equation.

- If $f(0) = 1$, then we will substitute $m \to -n$ into the original equation:

$$f(-n)f(n) = f(0)$$

from where

$$f(-n) = \frac{1}{f(n)}$$

for all integers n.

Let us now substitute $m \to 1$ into the original equation:

$$f(1)f(n) = f(1 + n)$$

Putting $f(1) = a$ we have that for all integers n

$$f(n + 1) = af(n)$$

For $a = 0$, we get $f(n + 1) = 0$ for all integer n and we have a constant zero function as in the previous case.

For $a \neq 0$, let us prove the following claim by induction on n.

Claim

Let $f : \mathbb{N} \to \mathbb{Z}$, such that $f(1) = a$, where $a \in \mathbb{Z}$. Given that for all $n \in \mathbb{N}$:

$$f(n + 1) = af(n)$$

then $f(n) = a^n$ for all $n \in \mathbb{N}$.

Proof

- **Basis of Induction**. The claim is true for $n = 1$, since $f(1) = a$.

- **Inductive Step**. Let us assume that the claim holds for $n = k$, i.e. that $f(k) = a^k$. Let us substitute $n \to k$ into the functional equation:

$$f(k + 1) = af(k)$$
$$f(k + 1) = a \cdot a^k$$
$$f(k + 1) = a^{k+1}$$

and the claim holds for $n = k + 1$. ∎

For the negative values of n we have

$$f(-n) = \frac{1}{f(n)} = \frac{1}{a^n} = a^{-n}$$

From here $f(n) = a^n$ for all integer n.

It is not hard to check that this function satisfies the original equation.

CHAPTER 41

Functional Equations on Rational Domain

In this chapter we will focus on the functional equations on rational domain. Induction is very useful for dealing with positive integer arguments. The case of zero is usually dealt apart. For negative integer arguments we need to establish either the even/odd properties or some similar identities. Once the case of the integer arguments is solved, then it is recommended to extend the solution to all rational numbers.

It is important to highlight that once we find a function that is a solution, we need to substitute it into the original function equation and verify if it indeed satisfies the conditions of the problem.

Let us consider several problems.

Problem 1

Find all functions $f : \mathbb{Q} \to \mathbb{R}$, such that for all $x, y \in \mathbb{Q}$

$$f(x + y) = f(x) + f(y)$$

Solution

Let us start by substituting $x \to 0$ and $y \to 0$ into the original equation:

$$f(0 + 0) = f(0) + f(0)$$
$$f(0) = 0$$

Substituting $y \to -x$ into the original equation we have

$$f(-x + x) = f(-x) + f(x)$$
$$f(0) = f(-x) + f(x)$$
$$0 = f(-x) + f(x)$$
$$-f(x) = f(-x)$$

and the function $f(x)$ is odd.

Let us now prove the following lemma by induction on n.

Lemma

If $f : \mathbb{Q} \to \mathbb{R}$ and for all $x, y \in \mathbb{Q}$:

$$f(x + y) = f(x) + f(y)$$

then

$$f(x_1 + x_2 + \ldots + x_n) = f(x_1) + f(x_2) + \ldots + f(x_n)$$

for all $x_i \in \mathbb{Q}$, $n \in \mathbb{N}$.

Proof

- **Basis of Induction**. The lemma is true for $n = 2$, since

$$f(x_1 + x_2) = f(x_1) + f(x_2)$$

- **Inductive Step**. Let us assume that the lemma holds for $n = k$, i.e. that

$$f(x_1 + x_2 + \ldots + x_k) = f(x_1) + f(x_2) + \ldots + f(x_k)$$

Let us make the following substitutions:

$$t = x_k + x_{k+1}$$
$$s = x_1 + x_2 + \ldots + x_k + x_{k+1}$$

Let us consider $f(s)$ and apply the inductive hypothesis for the variables $x_1, x_2, \ldots, t$:

$$
\begin{aligned}
f(s) &= f\left(x_1 + x_2 + \ldots + x_k + x_{k+1}\right) \\
&= f\left(x_1 + x_2 + \ldots + t\right) \\
&= f\left(x_1\right) + f\left(x_2\right) + \ldots + f\left(t\right) \\
&= f\left(x_1\right) + f\left(x_2\right) + \ldots + f\left(x_k + x_{k+1}\right) \\
&= f\left(x_1\right) + f\left(x_2\right) + \ldots + f\left(x_k\right) + f\left(x_{k+1}\right)
\end{aligned}
$$

and the lemma holds for $n = k + 1$. ∎

Let us put $f(1) = a$. We will prove that for all rational numbers x the solution of the equation is the linear function $f(x) = ax$.

Let us substitute $x_1 = x_2 = \ldots = x_n = 1$ into the lemma:

$$
\begin{aligned}
f\left(1 + 1 + \ldots + 1\right) &= f(1) + f(1) + \ldots + f(1) \\
f(n) &= nf(1) \\
f(n) &= an
\end{aligned}
$$

Substituting n for x we have that $f(x) = ax$ for all positive integers x.

Let n be a positive integer. Since the function is odd, then

$$
f(-n) = -f(n) = -an = a(-n)
$$

Substituting $-n$ for x we have that $f(x) = ax$ for all negative integers x.

Let us now substitute $x_1 = x_2 = \ldots = x_n = \frac{m}{n}$ into the lemma, where m and n are integers and $n \neq 0$:

$$
\begin{aligned}
f\left(\frac{m}{n} + \frac{m}{n} + \ldots + \frac{m}{n}\right) &= f\left(\frac{m}{n}\right) + f\left(\frac{m}{n}\right) + \ldots + f\left(\frac{m}{n}\right) \\
f\left(\frac{m}{n} \cdot n\right) &= f\left(\frac{m}{n}\right) \cdot n \\
f(m) &= f\left(\frac{m}{n}\right) \cdot n \\
am &= f\left(\frac{m}{n}\right) \cdot n \\
a\left(\frac{m}{n}\right) &= f\left(\frac{m}{n}\right)
\end{aligned}
$$

Substituting $\frac{m}{n}$ for x we have that $f(x) = ax$ for all rational numbers x.

Problem 2

Find all functions $f : \mathbb{Q} \to \mathbb{Q}^+$, such that $f(1) = 3$ and for all $m, n \in \mathbb{Q}$

$$f(x)f(y) = f(x + y)$$

Solution

Let us start by substituting $x \to 0$ and $y \to 0$ into the original equation:

$$f(0)f(0) = f(0 + 0)$$
$$f(0)f(0) = f(0)$$
$$f(0)\,(f(0) - 1) = 0$$

Since the codomain of the function $f(x)$ is $\mathbb{Q}^+$, then $f(0) = 1$.

Let us substitute $y \to -x$ into the original equation:

$$f(-x)f(x) = f(0)$$
$$f(-x)f(x) = 1$$
$$f(-x) = \frac{1}{f(x)}$$

Let us now prove the following lemma by induction on n.

Lemma

If $f : \mathbb{Q} \to \mathbb{Q}^+$ and for all $x, y \in \mathbb{Q}$:

$$f(x + y) = f(x)f(y)$$

then
$$f(x_1 + x_2 + \ldots + x_n) = f(x_1)\, f(x_2) \ldots f(x_n)$$

for all $x_i \in \mathbb{Q}$, $n \in \mathbb{N}$.

Proof

- **Basis of Induction**. The lemma is true for $n = 2$, since

$$f(x_1 + x_2) = f(x_1)\, f(x_2)$$

- **Inductive Step**. Let us assume that the lemma holds for $n = k$, i.e. that

$$f(x_1 + x_2 + \ldots + x_k) = f(x_1) f(x_2) \ldots f(x_k)$$

Let us make the following substitutions:

$$t = x_k + x_{k+1}$$
$$s = x_1 + x_2 + \ldots + x_k + x_{k+1}$$

Let us consider $f(s)$ and apply the inductive hypothesis for the variables $x_1, x_2, \ldots, t$:

$$\begin{aligned} f(s) &= f(x_1 + x_2 + \ldots + x_k + x_{k+1}) \\ &= f(x_1 + x_2 + \ldots + t) \\ &= f(x_1) f(x_2) \ldots f(t) \\ &= f(x_1) f(x_2) \ldots f(x_k + x_{k+1}) \\ &= f(x_1) f(x_2) \ldots f(x_k) f(x_{k+1}) \end{aligned}$$

and the lemma holds for $n = k + 1$. ∎

We will prove that for all rational numbers x the solution of the equation represents the exponential function $f(x) = 3^x$.

Let us first substitute $x_1 = x_2 = \ldots = x_n = 1$ into the lemma:

$$f(1)f(1) \ldots f(1) = f(1 + 1 + \ldots + 1)$$
$$3 \cdot 3 \cdot \ldots \cdot 3 = f(1 + 1 + \ldots + 1)$$
$$3^n = f(n)$$

Substituting n for x we have that $f(x) = 3^x$ for all positive integers x.

Let n be a positive integer. Then

$$f(-n) = \frac{1}{f(n)} = \frac{1}{3^n} = 3^{-n}$$

Substituting $-n$ for x we have that $f(x) = 3^x$ for all negative integers x.

Let us now substitute $x_1 = x_2 = \ldots = x_n = \frac{m}{n}$ into the lemma, where m and n are integers and $n \neq 0$:

$$f\left(\frac{m}{n} + \frac{m}{n} + \ldots + \frac{m}{n}\right) = f\left(\frac{m}{n}\right) f\left(\frac{m}{n}\right) \ldots f\left(\frac{m}{n}\right)$$

$$f\left(\frac{m}{n} \cdot n\right) = \left(f\left(\frac{m}{n}\right)\right)^n$$

$$f(m) = \left(f\left(\frac{m}{n}\right)\right)^n$$

$$3^m = \left(f\left(\frac{m}{n}\right)\right)^n$$

From the last equation we have

$$f\left(\frac{m}{n}\right) = 3^{\frac{m}{n}}$$

Substituting $\frac{m}{n}$ for x we have that $f(x) = 3^x$ for all rational numbers x.

Problem 3

Find all functions $f : \mathbb{Q} \to \mathbb{Q}$, such that $f(1) = 2$ and for all $m, n \in \mathbb{Q}$

$$f(xy) = f(x)f(y) - f(x + y) + 1$$

Solution

Let us make the following substitution

$$f(x) = g(x) + 1$$

Notice that the function $g(x)$ satisfies the initial condition $g(1) = 1$ and we can obtain the functional equation for $g(x)$:

$$g(xy) + 1 = (g(x) + 1)(g(y) + 1) - (g(x + y) + 1) + 1$$
$$g(xy) + 1 = g(x)g(y) + g(x) + g(y) + 1 - g(x + y) - 1 + 1$$
$$g(xy) = g(x)g(y) + g(x) + g(y) - g(x + y)$$

Let us substitute $x \to 0$ and $y \to 0$ into the functional equation for $g(x)$:

$$g(0) = g(0)g(0) + g(0) + g(0) - g(0 + 0)$$
$$0 = g^2(0)$$
$$0 = g(0)$$

Let us substitute $x \to 1$ and $y \to -1$ into the new functional equation:

$$g(-1) = g(1)g(-1) + g(1) + g(-1) - g(0)$$
$$g(-1) = g(-1) + 1 + g(-1)$$
$$-1 = g(-1)$$

Let us substitute $y \to 1$ into the new functional equation:

$$g(x) = g(x)g(1) + g(x) + g(1) - g(x+1)$$
$$g(x) = g(x) + g(x) + 1 - g(x+1)$$
$$g(x+1) = g(x) + 1$$

Let us substitute $x \to x - 1$ in the last equation:

$$g(x) = g(x-1) + 1$$

Now let us substitute $y \to -1$ into the functional equation for $g(x)$:

$$g(-x) = g(x)g(-1) + g(x) + g(-1) - g(x-1)$$
$$g(-x) = -g(x) + g(x) - 1 - g(x-1)$$
$$g(x-1) = -1 - g(-x)$$

Substituting this expression for $g(x-1)$ into the equation $g(x) = g(x-1) + 1$ we have

$$g(x) = -g(-x)$$

and the function $g(x)$ is odd.

Let us now prove the following lemma by induction on n.

Lemma 1

If $g : \mathbb{Q} \to \mathbb{Q}$ and $g(1) = 1$, and for all $x \in \mathbb{Q}$:

$$g(x+1) = g(x) + 1$$

then $g(n) = n$ for all $n \in \mathbb{N}$.

Proof

- **Basis of Induction**. The lemma is true for $n = 1$, since $g(1) = 1$.

- **Inductive Step**. Let us assume that the lemma holds for $n = k$, i.e. that $g(k) = k$. Let us substitute $x \to k$ into the functional equation:

$$g(k+1) = g(k) + 1 = k + 1$$

and the lemma holds for $n = k + 1$. ∎

Let n be a positive integer. Since the function $g(x)$ is odd, then

$$g(-n) = -g(n) = -n$$

Substituting $-n$ for x we have that $g(x) = x$ for all negative integers x.

Let us now prove the following lemma by induction on n.

Lemma 2

If $g : \mathbb{Q} \to \mathbb{Q}$ and for all $x \in \mathbb{Q}$:

$$g(x + 1) = g(x) + 1$$

then $g(x + n) = g(x) + n$ for all $x \in \mathbb{Q}$, $n \in \mathbb{N}$.

Proof

- **Basis of Induction.** The lemma is true for $n = 1$, since

$$g(x + 1) = g(x) + 1$$

- **Inductive Step.** Let us assume that the lemma holds for $n = k$, i.e. that $g(x + k) = g(x) + k$. Let us substitute $x \to k$ into the functional equation:

$$g(x + k + 1) = g(x + k) + 1 = g(x) + k + 1$$

and the lemma holds for $n = k + 1$.   ∎

Let n be a positive integer. Let us substitute $x \to x - n$ into $g(x + n) = g(x) + n$. We have

$$g(x) = g(x - n) + n$$
$$g(x) - n = g(x - n)$$

From here we conclude that

$$g(x + n) = g(x) + n$$

for all integer n.

Now let us substitute $x \to \frac{m}{n}$ and $y \to n$ into the functional equation for $g(x)$.

We have

$$g\left(\frac{m}{n} \cdot n\right) = g\left(\frac{m}{n}\right)g(n) + g\left(\frac{m}{n}\right) + g(n) - g\left(\frac{m}{n} + n\right)$$

$$g(m) = g\left(\frac{m}{n}\right)n + g\left(\frac{m}{n}\right) + n - g\left(\frac{m}{n}\right) - n$$

$$m = g\left(\frac{m}{n}\right)n$$

$$\frac{m}{n} = g\left(\frac{m}{n}\right)$$

Substituting $\frac{m}{n}$ for x we have that $g(x) = x$ for all rational numbers x.

This in turn implies that $f(x) = x + 1$ for all rational numbers x.

CHAPTER 42

Cauchy Functional Equation

Cauchy Functional Equation

The functional equation of the form

$$f(x + y) = f(x) + f(y)$$

is called **Cauchy Functional Equation.**

In Problem 1 of Chapter 41 "Functional Equations on Rational Domain" we solved the **Cauchy Functional Equation** on the set of rational numbers. We showed that the solution is a linear function of the form $f(x) = ax$, where a is some fixed real number. In this chapter we will focus on the problems that rely on the solutions of the **Cauchy Functional Equation** on real domain under the assumption that the function $f(x)$ is continuous.

Continuity

A function $f : A \to \mathbb{R}$, defined on a subset A of the real numbers, is said to be **continuous** at a point $x_0 \in A$ if for every sequence x_n in A that converges to x_0, the sequence $f(x_n)$ converges to $f(x_0)$.

Cauchy Functional Equation on Real Domain

Let $f : \mathbb{R} \to \mathbb{R}$ be a continuous function satisfying the Cauchy Functional Equation

$$f(x + y) = f(x) + f(y)$$

for all $x, y \in \mathbb{R}$. Then there exists a constant $a \in \mathbb{R}$, such that $f(x) = ax$ for all $x \in \mathbb{R}$.

Let us consider several problems.

Problem 1

Find all continuous functions $f : \mathbb{R} \to (0, \infty)$, such that for all $x, y \in \mathbb{R}$:

$$f(x + y) = f(x)f(y)$$

Solution

Since $f(x) > 0$, then let us take natural logarithm of both sides of the functional equation and rewrite it as follows:

$$f(x + y) = f(x)f(y)$$
$$\ln f(x + y) = \ln(f(x)f(y))$$
$$\ln f(x + y) = \ln f(x) + \ln f(y)$$

Let us consider the function $g(x) = \ln f(x)$. Then the last equation can be written in terms of $g(x)$:

$$g(x + y) = g(x) + g(y)$$

From here we see that $g(x)$ satisfies Cauchy Functional Equation on $\mathbb{R}$. Since $f(x)$ is continuous on $\mathbb{R}$, then $g(x)$ is also continuous on $\mathbb{R}$, as a composition of two continuous functions. Therefore, $g(x) = ax$ for some $a \in \mathbb{R}$.

From here we have

$$g(x) = \ln f(x)$$
$$ax = \ln f(x)$$
$$e^{ax} = f(x)$$

It is not hard to see that all exponential functions of the form $f(x) = e^{ax}$ satisfy the conditions of the problem.

Problem 2

Find all continuous functions $f : (0, +\infty) \to \mathbb{R}$, such that for all $x, y \in (0, +\infty)$

$$f(xy) = f(x) + f(y)$$

Solution

Since $x, y > 0$, then let us make the substitutions:

$$x = e^a$$
$$y = e^b$$

where $a, b \in \mathbb{R}$. Let us now rewrite the functional equation as follows:

$$f(xy) = f(x) + f(y)$$
$$f\left(e^a \cdot e^b\right) = f\left(e^a\right) + f\left(e^b\right)$$
$$f\left(e^{a+b}\right) = f\left(e^a\right) + f\left(e^b\right)$$

Let us consider the function $g(x) = f\left(e^x\right)$. Then the last equation can be written in terms of $g(x)$:

$$g(a + b) = g(a) + g(b)$$

From here we see that $g(x)$ satisfies Cauchy Functional Equation on $\mathbb{R}$. Since $f(x)$ is continuous on $(0, +\infty)$, then $g(x)$ is continuous on $\mathbb{R}$, as a composition of two continuous functions. Therefore, $g(x) = cx$ for some $c \in \mathbb{R}$.

For each $x \in \mathbb{R}$, let us put $y = e^x$ or equivalently $x = \ln y$. Then we have

$$g(x) = f\left(e^x\right)$$
$$cx = f\left(e^x\right)$$
$$c \ln y = f(y)$$

It is not hard to see that all logarithmic functions of the form $f(x) = c \ln x$ satisfy the conditions of the problem.

Problem 3

Find all continuous functions $f : \mathbb{R} \to \mathbb{R}$, such that for all $x, y \in \mathbb{R}$:

$$f(x + y) = f(x) + f(y) + f(x)f(y)$$

Solution

Answer: $f(x) = ax - 1$, where $a \in \mathbb{R}$.

Let us rewrite the functional equation by factoring it using the Simon's Favorite Factoring Trick:

$$f(x + y) = f(x) + f(y) + f(x)f(y)$$
$$f(x + y) + 1 = f(x) + f(y) + f(x)f(y) + 1$$
$$f(x + y) + 1 = (f(x)f(y) + f(x)) + (f(y) + 1)$$
$$f(x + y) + 1 = f(x)(f(y) + 1) + (f(y) + 1)$$
$$f(x + y) + 1 = (f(y) + 1)(f(x) + 1)$$

Let us consider the function $g(x) = f(x) + 1$. Then the last equation can be written in terms of $g(x)$:

$$g(x + y) = g(x)g(y)$$

In the Problem 1 of this Chapter we showed that the only functions that satisfy this equation are the function of the form $g(x) = e^{ax}$. From here we have $f(x) = e^{ax} - 1$.

It is not hard to see that all functions of the form $f(x) = e^{ax} - 1$ satisfy the conditions of the problem.

CHAPTER 43

Constructions in Functional Equations

In this chapter, we will focus on constructing a solution for the provided functional equation. The problems presented in this chapter involve functional equations with iterated functions. These problems typically inquire whether a function exists that fulfills a specific functional equation. It is crucial to observe that we are not required to solve the entire equation and identify all such functions. Instead, our goal is to discover at least one instance of a function that meets the specified criteria or to demonstrate the process of obtaining such an example.

Let us consider several problems.

Problem 1

Let $f^{(n)} = f(f(\ldots f(x) \ldots))$, where f is applied n times. Is there a non-zero function $f(x)$, such that

$$f^{(2023)}(x) + f^{(2022)}(x) + \ldots + f^{(2)}(x) + f^{(1)}(x) = f^{(2024)}(x)$$

Solution

Answer: yes.

Let us consider the functions of the form $f(x) = mx$. We will prove that there exists a real number $m_0 \neq 0$, such that $f(x) = m_0 x$ satisfies the conditions of the problem.

Let us prove the following lemma.

Lemma

Given the function $f(x) = mx$. Then

$$f^{(n)} = m^n x$$

Proof

Let us prove the claim of the lemma by induction on n.

- **Basis of Induction.** The lemma holds for $n = 1$ since $f(x) = mx$.

- **Inductive Step.** Let us assume that the lemma holds for $n = k$, i.e.

$$f^{(k)} = m^k x$$

Then

$$f^{(k+1)} = f\left(f^k(x)\right) = f\left(m^k x\right) = m^{k+1} x$$

and the lemma holds for $n = k + 1$. ∎

From the lemma it follows that for $f(x) = mx$ the functional equation becomes

$$f^{(2023)}(x) + f^{(2022)}(x) + \ldots + f^{(2)}(x) + f^{(1)}(x) = f^{(2024)}(x)$$

$$m^{2023} x + m^{2022} x + \ldots + m^2 x + mx = m^{2024} x$$

$$-m^{2024} x + m^{2023} x + \ldots + m^2 x + mx = 0$$

$$-mx\left(m^{2023} - m^{2022} - \ldots - m - 1\right) = 0$$

Let us consider the polynomial

$$P(m) = m^{2023} - m^{2022} - \ldots - m - 1$$

Since the degree of $P(m)$ is odd, then by Corollary for Odd-Degree Polynomials[1] $P(m)$ has at least one real root m_0. Now it is enough to put $f(x) = m_0 x$, which will satisfy the conditions of the problem.

---

[1] You can find more information about this topic in Chapter 15 "Roots of Polynomials".

Problem 2

Let n be a given positive integer. Is there a real number a, and a function $f(x)$, such that for all $x \in \mathbb{R}$:

$$f(f(\ldots f(x)\ldots)) = (x+a)^{2^n} + 1$$

where the function f is applied n times.

Solution

Answer: yes.

Let us prove the following lemma.

Lemma

For the function $f(x) = (x-1)^2 + 1$ we have

$$f(f(\ldots f(x)\ldots)) = (x-1)^{2^n} + 1$$

where the function f is applied n times.

Proof

Let us prove the claim of the lemma by induction on n.

- **Basis of Induction**. The lemma holds for $n = 1$ since

$$f(x) = (x-1)^2 + 1$$

- **Inductive Step**. Let us assume that the lemma holds for $n = k$, i.e.

$$f(f(\ldots f(x)\ldots)) = (x-1)^{2^k} + 1$$

where the function f is applied k times. Then

$$f\left((x-1)^{2^k} + 1\right) = \left((x-1)^{2^k} + 1 - 1\right)^2 + 1$$

$$= \left((x-1)^{2^k}\right)^2 + 1$$

$$= (x-1)^{2^{k+1}} + 1$$

and the lemma holds for $n = k+1$. ∎

Now it is enough to put $a = -1$ and the function $f(x) = (x-1)^2 + 1$ will satisfy the conditions of the problem.

Problem 3

Let $f(x)$ be some function, such that $f : [0, +\infty) \to \mathbb{R}$. Let us define $g_1(x) = f(x)$ and $g_n = f(g_{n-1}(x))$. Is there $f(x)$, such that

$$g_{2023}(x) = 1 + x + 2\sqrt{x}$$

Solution

Answer: yes.

Let us start by proving the following lemma.

Lemma

Let $b \in \mathbb{R}^+$ and $f(x) = (b + \sqrt{x})^2$. Then

$$g_n(x) = (bn + \sqrt{x})^2$$

Proof

Let us prove the claim of the lemma by induction on n.

- **Basis of Induction**. The lemma holds for $n = 1$ since

$$g_1(x) = f(x) = (b + \sqrt{x})^2$$

- **Inductive Step**. Let us assume that the lemma holds for $n = k$, i.e.

$$g_k(x) = (bk + \sqrt{x})^2$$

Then

$$g_{k+1}(x) = f(g_k(x))$$
$$= f\left((bk + \sqrt{x})^2\right)$$
$$= \left(b + \sqrt{(bk + \sqrt{x})^2}\right)^2$$
$$= (b + bk + \sqrt{x})^2$$
$$= (b(k + 1) + \sqrt{x})^2$$

and the lemma holds for $n = k + 1$.   ∎

Now it is enough to put $b = \frac{1}{2023}$ and the function $f(x) = (b + \sqrt{x})^2$ will satisfy the conditions of the problem. Indeed, from the lemma we have

$$g_{2023}(x) = \left(2023b + \sqrt{x}\right)^2$$

$$= \left(2023 \cdot \frac{1}{2023} + \sqrt{x}\right)^2$$

$$= \left(1 + \sqrt{x}\right)^2$$

$$= 1 + x + 2\sqrt{x}$$

as desired.

CHAPTER 44

Sequences

Sequence

Mapping of the form

$$a : \mathbb{N} \to \mathbb{R}$$

is called **sequence** and is denoted as a_n.

Math olympiad algebra problems that involve sequences use either recursive or explicit formulas. It is customary to employ induction, inequalities, substitutions and algebraic manipulations in these types of problems.

Let us consider several problems.

Problem 1

Given a periodic sequence of real numbers a_n, such that for all integers $n \geq 2$

$$a_{n+1} = \frac{4}{5}\left(a_{n-1} + \frac{1}{a_n}\right)$$

Prove that

$$a_1 \cdot a_2 \cdot \ldots \cdot a_{2n-1} \cdot a_{2n} = 4^n$$

Solution

Let us rewrite the equation as follows:

$$a_{n+1} = \frac{4}{5}\left(a_{n-1} + \frac{1}{a_n}\right)$$

$$a_{n+1} = \frac{4}{5}\left(\frac{a_{n-1}a_n + 1}{a_n}\right)$$

$$a_n a_{n+1} = \frac{4}{5}\left(a_{n-1}a_n + 1\right)$$

$$a_n a_{n+1} - 4 = \frac{4}{5}\left(a_{n-1}a_n - 4\right)$$

Let us make the substitution $b_n = a_{n-1}a_n - 4$. Then the last equality becomes

$$b_{n+1} = \frac{4}{5}b_n$$

and b_n is a geometric sequence. Since the sequence a_n is periodic, then so is b_n. This implies that $b_n = 0$, and, therefore

$$a_{n-1}a_n = 4$$

for all $n \geq 2$. From here we have

$$a_1 \cdot a_2 \cdot \ldots \cdot a_{2n-1} \cdot a_{2n} = (a_1 a_2) \cdot \ldots \cdot (a_{2n-1}a_{2n}) = 4^n$$

as desired.

Problem 2

Given a sequence of integer numbers a_n, such that for all $n \geq 1$ the equation

$$a_{n+2}x^2 + a_{n+1}x + a_n = 0$$

has at least one real solution for x. Prove that the sequence a_n cannot have infinite number of terms.

Solution

Let D be the discriminant of the given quadratic equation. Let us consider the inequality $D \geq 0$ and rewrite it as follows:

$$D \geq 0$$

$$a_{n+1}^2 - 4a_n a_{n+2} \geq 0$$

$$a_{n+1}^2 \geq 4a_n a_{n+2}$$

$$\frac{1}{4} \cdot \frac{a_{n+1}}{a_n} \geq \frac{a_{n+2}}{a_{n+1}}$$

Let us make the substitution $b_n = \frac{a_{n+1}}{a_n}$. Then we have

$$b_{n+1} \leq \frac{1}{4} \cdot b_n \leq \frac{1}{4^2} \cdot b_{n-1} \leq \ldots \leq \frac{1}{4^n} \cdot b_1$$

Since $b_1 = \frac{a_2}{a_1}$, then b_1 is fixed and there exists a positive integer k, such that

$$\frac{1}{4^k} \cdot b_1 < 1$$

Indeed, this inequality is equivalent to

$$k > \log_4 b_1$$

and it is enough to take the ceiling function[1] for the value of k:

$$k = \lceil \log_4 b_1 \rceil$$

Then for all $n > k$ we have that $b_n < 1$. This implies that $a_{n+1} < a_n$ and the sequence a_n is strictly decreasing for all $n > k$. However, since any sequence of natural numbers cannot infinitely decrease, then the sequence a_n cannot have infinite number of terms.

Problem 3

Given the sequence of real numbers a_n, such that $a_1 = 1$ and

$$a_{n+1} = \frac{3a_n}{4} + \frac{1}{a_n}$$

for all integers $n \geq 1$. Prove that for all positive integers m, there exists a positive integer n, such that

$$ma_n > 2m - 1$$

---

[1] This function is discussed in detail in Chapter 46 "Ceiling Function".

Solution

Let us rewrite the inequality as

$$ma_n > 2m - 1$$
$$1 > 2m - ma_n$$
$$\frac{1}{m} > 2 - a_n$$

Let us prove the following lemma by induction on n.

Lemma

For all integers $n \geq 2$:

$$0 < 2 - a_n < \left(\frac{3}{4}\right)^n$$

Proof

- **Basis of Induction.** Let us show that the lemma holds for $n = 2$. We have $a_2 = \frac{7}{4}$, and therefore $2 - a_2 = \frac{1}{4}$. From here

$$0 < \frac{1}{4} = \frac{4}{16} < \frac{9}{16} < \left(\frac{3}{4}\right)^2$$

and the lemma holds for $n = 2$.

- **Inductive Step.** Let us assume that

$$0 < 2 - a_k < \left(\frac{3}{4}\right)^k$$

for some positive integer k and let us prove that

$$0 < 2 - a_{k+1} < \left(\frac{3}{4}\right)^{k+1}$$

Let us consider the expression for $2 - a_{k+1}$ and use the recursive formula given in the problem.

We have

$$2 - a_{k+1} = 2 - \left(\frac{3a_k}{4} + \frac{1}{a_k} \right)$$

$$= 2 - \frac{3a_k}{4} - \frac{1}{a_k}$$

$$= \frac{3}{4} \left(2 - a_k \right) + \frac{1}{2} - \frac{1}{a_k}$$

$$< \frac{3}{4} \left(\frac{3}{4} \right)^k + \frac{1}{2} - \frac{1}{a_k}$$

$$= \left(\frac{3}{4} \right)^{k+1} - \frac{2 - a_k}{2a_k}$$

$$< \left(\frac{3}{4} \right)^{k+1}$$

and the lemma is proven. ∎

Now it will be enough to prove that for all positive integers m, there exists a positive integer n, such that

$$\left(\frac{3}{4} \right)^n < \frac{1}{m}$$

Let us rewrite this inequality as follows:

$$\frac{1}{m} > \left(\frac{3}{4} \right)^n$$

$$\left(\frac{4}{3} \right)^n > m$$

$$\ln \left(\frac{4}{3} \right)^n > \ln m$$

$$n \ln \left(\frac{4}{3} \right) > \ln m$$

$$n > \frac{\ln m}{\ln \left(\frac{4}{3} \right)}$$

For the needed value of n it is enough to take the ceiling function[2] and put

$$n = \left\lceil \frac{\ln m}{\ln \left(\frac{4}{3} \right)} \right\rceil$$

[2]This function is discussed in detail in Chapter 46 "Ceiling Function".

CHAPTER 45

Floor Function

The **floor function** is a mathematical function that rounds a real number down to the nearest integer less than or equal to that number. In other words, it gives back the greatest integer that is less than or equal to the given real number. The **floor function** proves highly valuable in problems that straddle the boundary between algebra and number theory.

Floor Function

Floor function $\lfloor x \rfloor$ of the real number x is defined as

$$\lfloor x \rfloor = n$$

where n is an integer number, such that $n \leq x < n + 1$.

Properties of Floor Function

1. For integer numbers n:
$$\lfloor n \rfloor = n$$

2. For real number x:
$$x - 1 < \lfloor x \rfloor \leq x$$

3. For real number x and integer number n:
$$\lfloor x + n \rfloor = \lfloor x \rfloor + n$$

4. For real numbers x and y:
$$\lfloor x \rfloor + \lfloor y \rfloor \leq \lfloor x + y \rfloor \leq \lfloor x \rfloor + \lfloor y \rfloor + 1$$

5. The floor function is nondecreasing, i.e. for real numbers $x \leq y$:
$$\lfloor x \rfloor \leq \lfloor y \rfloor$$

Let us now consider several problems that involve the floor function.

Problem 1

Prove that for all real numbers $a > 1$:
$$\left\lfloor \sqrt{\lfloor \sqrt{a} \rfloor} \right\rfloor = \left\lfloor \sqrt{\sqrt{a}} \right\rfloor$$

Solution

Notice that for all $a > 1$ there exist a positive integer n, such that
$$n^4 \leq a < (n+1)^4$$

Let us prove that
$$\left\lfloor \sqrt{\lfloor \sqrt{a} \rfloor} \right\rfloor = \left\lfloor \sqrt{\sqrt{a}} \right\rfloor = n$$

Taking the square root of the three sides of the above inequality we have
$$n^2 \leq \sqrt{a} < (n+1)^2$$

Then
$$n^2 \leq \lfloor \sqrt{a} \rfloor < (n+1)^2$$
and taking the square root of the three sides of this inequality we have
$$n \leq \sqrt{\lfloor \sqrt{a} \rfloor} < n+1$$
From the definition of the floor function we have
$$\sqrt{\lfloor \sqrt{a} \rfloor} = n$$
However, considering
$$n^2 \leq \sqrt{a} < (n+1)^2$$
and taking the square root of the three sides of this inequality we have
$$n \leq \sqrt{\sqrt{a}} < n+1$$
This implies that
$$\left\lfloor \sqrt{\sqrt{a}} \right\rfloor = n$$
as desired.

Problem 2

Are there irrational numbers $\alpha, \beta > 1$, such that $\lfloor \alpha^m \rfloor \neq \lfloor \beta^n \rfloor$ for all $m, n \in \mathbb{N}$?

Solution

Answer: yes.

Let us put $\alpha = 3\sqrt{3}$ and $\beta = 2\sqrt{2}$ and prove that $|\alpha^m - \beta^n| > 1$ for all $m, n \in \mathbb{N}$. This will automatically imply that $\lfloor \alpha^m \rfloor \neq \lfloor \beta^n \rfloor$ for all $m, n \in \mathbb{N}$.

Start by noticing that for all real numbers $x > y > 1$:
$$x^3 - y^3 > x^2 - y^2$$
Indeed, this inequality can be equivalently transformed as follows
$$x^3 - y^3 - x^2 + y^2 > 0$$
$$(x-y)\left(x^2 + xy + y^2\right) - (x-y)(x+y) > 0$$
$$(x-y)\left(x^2 + xy + y^2 - x - y\right) > 0$$
$$(x-y)\left(2x^2 + 2xy + 2y^2 - 2x - 2y\right) > 0$$
$$(x-y)\left((x+y)^2 + (x-1)^2 + (y-1)^2 - 2\right) > 0$$

which obviously holds for $x > y > 1$.

Now we have

$$|\alpha^m - \beta^n| = \left|\left(\sqrt{3m}\right)^3 - \left(\sqrt{2n}\right)^3\right| > |3^m - 2^n| \geq 1$$

as desired.

Problem 3

A real nonzero number x satisfies the equation

$$\lfloor x \rfloor^3 + \{x\}^3 + 3x = x^3$$

Prove that x is a rational number greater than 2.

Solution

We will solve this problem by doing the following casework:

- If x is an integer, then let us put $x = n$. Therefore, the fractional part[1] of n is zero and the equation becomes

$$n^3 + 3n = n^3$$

  This implies that $n = 0$ and we obtained a contradiction.

- If x is not an integer, then let us put $\lfloor x \rfloor = n$ and $\{x\} = a$, where $a \neq 0$. Then $x = n + a$ and the equation becomes

$$n^3 + a^3 + 3(n + a) = (n + a)^3$$
$$n^3 + a^3 + 3(n + a) = n^3 + 3n^2a + 3na^2 + a^3$$
$$3(n + a) = 3n^2a + 3na^2$$
$$3(n + a) = 3na(n + a)$$

Since x is nonzero, then $n + a \neq 0$ and from the last equation we have that $1 = na$. Notice that the fractional part is always nonnegative. Then n is a positive integer and

$$\{x\} = a = \frac{1}{n}$$

For $n = 1$ we have $\{x\} = 1$ and we obtained a contradiction.

---

[1] This function is discussed in detail in Chapter 47 "Fractional Part".

For $n \geq 2$ we have that

$$x = \lfloor x \rfloor + \{x\} = n + \frac{1}{n}$$

which implies that x is a rational number.

Now from the AM-GM Inequality we have

$$x = n + \frac{1}{n} \geq 2\sqrt{n \cdot \frac{1}{n}} = 2$$

Since in our case the equality is not reached, then x is strictly greater than 2 as desired.

CHAPTER 46

Ceiling Function

The **ceiling function** is a mathematical function that rounds a real number up to the nearest integer greater than or equal to that number. In other words, it gives back the least integer that is greater than or equal to the given real number. The **ceiling function** proves highly valuable in problems that straddle the boundary between algebra and number theory.

Ceiling Function

Ceiling function $\lceil x \rceil$ of the real number x is defined as

$$\lceil x \rceil = n$$

where n is an integer number, such that $n - 1 < x \leq n$.

Properties of Ceiling Function

1. For integer numbers n:
$$\lceil n \rceil = n$$

2. For real number x:
$$x \leq \lceil x \rceil < x + 1$$

3. For real number x and integer number n:
$$\lceil x + n \rceil = \lceil x \rceil + n$$

4. For real numbers x and y:
$$\lceil x \rceil + \lceil y \rceil - 1 \leq \lceil x + y \rceil \leq \lceil x \rceil + \lceil y \rceil$$

5. The ceiling function is nondecreasing, i.e. for real numbers $x \leq y$:
$$\lceil x \rceil \leq \lceil y \rceil$$

Let us now consider several problems that involve the ceiling function.

Problem 1

The sequence is given as $a_1 = 1$ and for all integers $n \geq 1$

$$a_{n+1} = \lceil a_n \rceil + \frac{1}{a_n}$$

Prove that there exists a positive integer k, such that $\lceil a_k \rceil = 2023$.

Solution

Start by noticing that

$$a_2 = 2$$
$$a_3 = 2.5$$

Let us now prove the following lemma.

Lemma

For all integers $n \geq 3$:
$$n - 1 < a_n < n$$

Proof

We will prove the claim of the lemma by induction on n.

- **Basis of Induction.** For $n = 3$ the lemma obviously holds:

$$2 < a_3 < 3$$

- **Inductive Step.** Let us assume that the lemma holds for $n = k$, i.e. that $k - 1 < a_k < k$, and prove that the it also holds for $n = k + 1$:

$$k < a_{k+1} < k + 1$$

Notice that by the inductive hypothesis $\lceil a_k \rceil = k$. Then

$$a_{k+1} = \lceil a_k \rceil + \frac{1}{a_k} = k + \frac{1}{a_k} > k + \frac{1}{k} > k$$

and also

$$a_{k+1} = \lceil a_k \rceil + \frac{1}{a_k} = k + \frac{1}{a_k} < k + \frac{1}{k-1} < k + 1$$

This implies that the lemma holds for $n = k + 1$ and the lemma is proven. ∎

Now it is enough to apply the lemma for $k = 2023$. Indeed, we have

$$2022 < a_{2023} < 2023$$

which implies that $\lceil a_k \rceil = 2023$, as desired.

Problem 2

Solve the equation in real numbers

$$x^3 - \lceil x \rceil = 3$$

Solution

Let us put $\lceil x \rceil = n$. Then
$$x + 1 - \{x\} = n$$
where $\{x\}$ is the fractional part[1] of x. Now the equation becomes
$$x^3 - (x + 1 - \{x\}) = 3$$
$$x^3 - x - 1 + \{x\} = 3$$
$$\{x\} = -x^3 + x + 4$$

Since $0 \le \{x\} < 1$, then
$$0 \le -x^3 + x + 4 < 1$$

Let us consider the function
$$f(t) = -t^3 + t + 4$$

It's first derivative is
$$f'(t) = -3t^2 + 1$$

and it's critical points are $t = \pm\frac{\sqrt{3}}{3}$. The function is decreasing on $\left(-\infty, -\frac{\sqrt{3}}{3}\right]$ and increasing on $\left[-\frac{\sqrt{3}}{3}, \frac{\sqrt{3}}{3}\right]$. Therefore, for $t \in \left(-\infty, \frac{\sqrt{3}}{3}\right]$ we have

$$f(t) \ge f\left(-\frac{\sqrt{3}}{3}\right) = 4 - \frac{2\sqrt{3}}{9} > 1$$

This implies that there are no solutions on the interval $\left(-\infty, \frac{\sqrt{3}}{3}\right]$.

The function is decreasing on $\left[\frac{\sqrt{3}}{3}, \infty\right)$, and, therefore, for $t \ge 2$ we have

$$f(t) \le f(2) = -2 < 0$$

This implies that there are no solutions on the interval $[2, \infty)$.

Therefore, we only need to check the interval $\left(\frac{\sqrt{3}}{3}, 2\right)$. Notice that

$$0 < \frac{\sqrt{3}}{3} < 1$$

If $x \in \left(\frac{\sqrt{3}}{3}, 2\right)$, then $\lceil x \rceil$ can only be equal to 1 or 2.

If $\lceil x \rceil = 1$, then the equation becomes

$$x^3 - 1 = 3$$

---

[1] This function is discussed in detail in Chapter 47 "Fractional Part".

which implies that $x = \sqrt[3]{4}$. However, since $\sqrt[3]{4} > 1$, then $\lceil x \rceil = 2$ and we obtained a contradiction.

If $\lceil x \rceil = 2$, then the equation becomes

$$x^3 - 2 = 3$$

which implies that $x = \sqrt[3]{5}$. It is not hard to check that this value of x indeed satisfies the initial equation

Problem 3

Show that the equation

$$\lceil x^2 \rceil - \lceil 2^x \rceil = \{2x\} - 2$$

has no solutions in real numbers.

Solution

Notice that the left-hand side of the equation is an integer and, therefore, so should be the right-hand side. This implies that the fractional part[2] $\{2x\}$ is zero and $2x$ is an integer.

We will proceed by doing the following casework:

- If x is an integer, then let us put $x = n$. The equation now becomes

$$n^2 - 2^n = -2$$
$$n^2 + 2 = 2^n$$

Notice that $n = 0, 1, 2, 3, 4$ are not the solutions of the last equation.

For $n \leq 0$ we have
$$2^n \leq 2^0 = 1 < 2 \leq n^2 + 2$$

For $n \geq 5$ we have
$$2^n > n^2 + 2$$

Indeed, by Bernoulli's Inequality[3]

$$2^n = 2^4 \cdot 2^{\frac{n}{2}-2} \cdot 2^{\frac{n}{2}-2} \geq 16\left(\frac{n}{2}-1\right)^2 = 4n^2 - 16n + 16 > n^2 + 2$$

and we conclude that there are no solutions in this case.

---

[2] This function is discussed in detail in Chapter 47 "Fractional Part".
[3] This inequality is discussed in detail in Chapter 27 "Bernoulli's Inequality".

- If x is not an integer, but $2x$ is an integer, then $x = \frac{2n+1}{2}$, where n is an integer. The equation now becomes

$$\left\lceil \left(\frac{2n+1}{2}\right)^2 \right\rceil - \left\lceil 2^{\frac{2n+1}{2}} \right\rceil = -2$$

$$\left\lceil \frac{4n^2 + 4n + 1}{4} \right\rceil - \left\lceil 2^{\frac{2n+1}{2}} \right\rceil = -2$$

$$\left\lceil n^2 + n + \frac{1}{4} \right\rceil - \left\lceil 2^{\frac{2n+1}{2}} \right\rceil = -2$$

$$n^2 + n + 1 - \left\lceil 2^{\frac{2n+1}{2}} \right\rceil = -2$$

$$n^2 + n + 3 = \left\lceil 2^{\frac{2n+1}{2}} \right\rceil$$

Notice that $n = 0, 1, 2, 3, 4$ are not the solutions of the last equation.

For $n \leq -1$ we have

$$\left\lceil 2^{\frac{2n+1}{2}} \right\rceil < 2^{\frac{2n+1}{2}} + 1 \leq 2^{-\frac{1}{2}} + 1 < 2 < 3 \leq n^2 + n + 3$$

For $n \geq 5$ we have

$$\left\lceil 2^{\frac{2n+1}{2}} \right\rceil \geq 2^{\frac{2n+1}{2}} > n^2 + n + 3$$

Indeed, by Bernoulli's Inequality

$$2^{\frac{2n+1}{2}} > 2^4 \cdot 2^{\frac{n}{2}-2} \cdot 2^{\frac{n}{2}-2} \geq 16 \left(\frac{n}{2} - 1\right)^2 = 4n^2 - 16n + 16 > n^2 + n + 3$$

and we conclude that there are no solutions in this case.

CHAPTER 47

Fractional Part

The **fractional part** is a mathematical function that extracts the decimal portion of a real number by removing its integer part. In other words, it returns the difference between the number and its floor function[1]. The **fractional part** proves highly valuable in problems that straddle the boundary between algebra and number theory.

Fractional Part

Fractional part $\{x\}$ of the real number x is defined as

$$\{x\} = x - \lfloor x \rfloor$$

where $\lfloor x \rfloor$ is the floor function.

---

[1]This function is discussed in detail in Chapter 45 "Floor Function".

Properties of Fractional Part

1. For integer numbers n:
$$\{n\} = 0$$

2. For real number x:
$$0 \leq \{x\} < 1$$

3. For real number x and integer number n:
$$\{x + n\} = \{x\}$$

4. For real noninteger numbers x:
$$\{-x\} = 1 - \{x\}$$

Let us now consider several problems that involve the fractional part.

Problem 1

Solve the equation in real numbers:
$$\{x\} \cdot \{y\} = \{x + y\}$$

Solution

We will solve this problem by doing the following casework:

- If any of the numbers x or y is integer, then without loss of generality we can assume that it is x. Let us put $x = n$. Then $\{n\} = 0$ and the equation becomes
$$\{n + y\} = 0$$

This implies that $\{y\} = 0$ and y is also integer. It is not hard to see that all pairs of integer numbers (x, y) satisfy the original equation.

- If none of the numbers x or y is integer, then let us put $\{x\} = a$ and $\{y\} = b$. Then $0 < a, b < 1$ and the equation becomes
$$ab = \{a + b\}$$

Notice that this implies that $ab - (a+b)$ is an integer number. Then the following number is also integer:
$$ab - (a + b) + 1 = ab - a - b + 1 = a(b - 1) - (b - 1) = (a - 1)(b - 1)$$

However, since $0 < a, b < 1$, we have that

$$0 < (a-1)(b-1) < 1$$

which leads to a contradiction.

We conclude that the solutions of the original equation are all pairs of integer numbers (x, y).

Problem 2

Given a positive integer number n. Prove that

$$\left\{\left(2+\sqrt{3}\right)^{2n}\right\} > 0.99\ldots99$$

where the digit 9 appears n times after the decimal point.

Solution

Let us put

$$A = \left(2+\sqrt{3}\right)^{2n}$$

$$B = \left(2-\sqrt{3}\right)^{2n}$$

Notice that A and B are conjugates and

$$A \cdot B = \left(2+\sqrt{3}\right)^{2n} \cdot \left(2-\sqrt{3}\right)^{2n} = 1$$

First, let us show that $\{B\} = B$. Indeed, we have

$$0 < 2-\sqrt{3} < 1$$

$$0 < \left(2-\sqrt{3}\right)^{2n} < 1$$

$$0 < B < 1$$

and, therefore, $\{B\} = B$.

From the Binomial Theorem it follows that after the expansion the number $A + B$ becomes integer. Therefore, we have

$$\{A\} + \{B\} = 1$$

$$\{A\} = 1 - \{B\}$$

$$\{A\} = 1 - B$$

Now let us prove that $B < 10^{-n}$. Indeed

$$B = \frac{1}{A} = \frac{1}{\left(2 + \sqrt{3}\right)^{2n}} = \frac{1}{\left(7 + 2\sqrt{3}\right)^{n}} < \frac{1}{10^{n}} = 10^{-n}$$

Then

$$\{A\} = 1 - B > 1 - 10^{-n} = 0.99\ldots 99$$

as desired.

Problem 3

Prove the inequality for all positive integers n:

$$\sum_{i=1}^{n^2} \left\{\sqrt{i}\right\} \leq \frac{n^2 - 1}{2}$$

Solution

We will solve this problem by doing the following casework:

- For $n = 1$ the inequality becomes equality:

$$\left\{\sqrt{1}\right\} = \frac{(1)^2 - 1}{2}$$

- For $n > 1$ let us prove the following lemma.

Lemma

$$\sum_{i=m^2}^{m^2 + 2m} \left\{\sqrt{i}\right\} \leq \frac{2m + 1}{2}$$

Proof

Let us put

$$F_m(a) = \left\{\sqrt{m^2 + a}\right\} + \left\{\sqrt{m^2 + 2m - a}\right\}$$

From the AM-QM Inequality we have

$$\sqrt{m^2 + a} + \sqrt{m^2 + 2m - a} \leq \sqrt{2(2m^2 + 2m)}$$
$$= \sqrt{4m^2 + 4m}$$
$$< \sqrt{4m^2 + 4m + 1}$$
$$= 2m + 1$$

Therefore

$$F_m(a) = \left\{\sqrt{m^2 + a}\right\} + \left\{\sqrt{m^2 + 2m - a}\right\} \le 1$$

Adding the last inequality for $a = 0, 1, \ldots, m$ we have

$$\sum_{i=m^2}^{m^2+2m} \left\{\sqrt{i}\right\} = \left\{\sqrt{m^2 + m}\right\} + \sum_{a=0}^{m-1} F_m(a)$$

$$\le \frac{1}{2} + \sum_{a=0}^{m-1} 1$$

$$= \frac{1}{2} + m$$

$$= \frac{2m + 1}{2}$$

and the lemma is proven. ∎

Using the above lemma for $m = 1, 2, \ldots, n - 1$ we have

$$\sum_{i=1}^{n^2} \left\{\sqrt{i}\right\} \le \sum_{m=1}^{n-1} \frac{2m + 1}{2}$$

$$= \sum_{m=1}^{n-1} m + \frac{1}{2} \sum_{m=1}^{n-1} 1$$

$$= \frac{(n - 1)n}{2} + \frac{n - 1}{2}$$

$$= \frac{n^2 - 1}{2}$$

which is what needed to be proven.

CHAPTER 48

Trigonometric Substitution

Trigonometric Substitution is a technique employed to solve math olympiad problems by substituting the given variables with the trigonometric functions of new variables. The trigonometric functions are typically selected based on the values taken by the initial variables and should align with the range of the corresponding trigonometric functions. After the substitution, we commonly utilize various trigonometric identities and perform algebraic manipulations. This technique becomes exceedingly valuable when we need to verify identities or prove inequalities that would be exceptionally challenging to accomplish using alternative methods.

Typical Trigonometric Substitutions

- If a is any real number, then a can be substituted for $\tan x$ or $\cot x$.

- If $|a| \leq 1$, then a can be substituted for $\sin x$ or $\cos x$.

- If $|a| \geq 1$, then a can be substituted for $\csc x$ or $\sec x$.

Trigonometric Identities

- Pythagorean Identities

$$\sin^2 x + \cos^2 x = 1$$
$$1 + \cot^2 x = \csc^2 x$$
$$1 + \tan^2 x = \sec^2 x$$

- Angle Addition Identities

$$\sin(x + y) = \sin x \cos y + \cos x \sin y$$
$$\cos(x + y) = \cos x \cos y - \sin x \sin y$$
$$\tan(x + y) = \frac{\tan x + \tan y}{1 - \tan x \tan y}$$

- Angle Subtraction Identities

$$\sin(x - y) = \sin x \cos y - \cos x \sin y$$
$$\cos(x - y) = \cos x \cos y + \sin x \sin y$$
$$\tan(x - y) = \frac{\tan x - \tan y}{1 + \tan x \tan y}$$

- Double-Angle Identities

$$\sin(2x) = 2 \sin x \cos x$$
$$\cos(2x) = \cos^2 x - \sin^2 x$$
$$\cos(2x) = 1 - 2\sin^2 x$$
$$\cos(2x) = 2\cos^2 x - 1$$
$$\tan(2x) = \frac{2\tan x}{1 - \tan^2 x}$$

- Triple-Angle Identities

$$\sin(3x) = 3\sin x - 4\sin^3 x$$
$$\cos(3x) = 4\cos^3 x - 3\cos x$$
$$\tan(3x) = \frac{3\tan x - \tan^3 x}{1 - 3\tan^2 x}$$

- Half-Angle Identities

$$\sin\left(\frac{x}{2}\right) = \pm\sqrt{\frac{1 - \cos x}{2}}$$

$$\cos\left(\frac{x}{2}\right) = \pm\sqrt{\frac{1 + \cos x}{2}}$$

$$\tan\left(\frac{x}{2}\right) = \pm\sqrt{\frac{1 - \cos x}{1 + \cos x}}$$

$$\tan\left(\frac{x}{2}\right) = \frac{\sin x}{1 + \cos x}$$

$$\tan\left(\frac{x}{2}\right) = \frac{1 - \cos x}{\sin x}$$

Let us consider several problems.

Problem 1

Find all positive integer n, such that for all $|x| \leq 1$ it holds that

$$(1 + x)^n + (1 - x)^n \leq 2^n$$

Solution

We will solve this problem by doing the following casework:

- If $n = 1$, then the inequality becomes

$$(1 + x) + (1 - x) \leq 2$$
$$2 \leq 2$$

which obviously holds.

- If $n \geq 2$, then let us make the substitution $x = \cos\alpha$, where $\alpha \in [0, 2\pi)$.

We have

$$(1 + x)^n + (1 - x)^n = (1 + \cos \alpha)^n + (1 - \cos \alpha)^n$$

$$= \left(2 \cdot \frac{1 + \cos \alpha}{2} \right)^n + \left(2 \cdot \frac{1 - \cos \alpha}{2} \right)^n$$

$$= 2^n \cdot \left(\frac{1 + \cos \alpha}{2} \right)^n + 2^n \cdot \left(\frac{1 - \cos \alpha}{2} \right)^n$$

$$= 2^n \cdot \cos^n \left(\frac{\alpha}{2} \right) + 2^n \cdot \sin^n \left(\frac{\alpha}{2} \right)$$

$$= 2^n \cdot \left(\cos^n \left(\frac{\alpha}{2} \right) + \sin^n \left(\frac{\alpha}{2} \right) \right)$$

$$\leq 2^n \cdot \left(\cos^2 \left(\frac{\alpha}{2} \right) + \sin^2 \left(\frac{\alpha}{2} \right) \right)$$

$$= 2^n$$

which is what needed to be proven.

Problem 2

Is it true that that among any 7 real numbers x_1, x_2, ..., x_7, there always exist two numbers x_i and x_j, such that

$$0 < \frac{x_i - x_j}{1 + x_i x_j} < \frac{\sqrt{3}}{3}$$

Solution

Let us put $x_i = \tan \alpha_i$ for $i = 1, 2, \ldots, 7$, where $\alpha_i \in \left(-\frac{\pi}{2}, \frac{\pi}{2} \right)$. Then the inequality becomes

$$0 < \frac{\tan \alpha_i - \tan \alpha_j}{1 + \tan \alpha_i \tan \alpha_j} < \frac{\sqrt{3}}{3}$$

$$0 < \tan (\alpha_i - \alpha_j) < \frac{\sqrt{3}}{3}$$

Notice that if we divide the interval $\left(-\frac{\pi}{2}, \frac{\pi}{2} \right)$ into 6 equal segments of length $\frac{\pi}{6}$ each, then by the Pigeonhole Principle there exist two numbers α_i and α_j that belong to the same segment. Therefore, we have

$$0 < \alpha_i - \alpha_j < \frac{\pi}{6}$$

and

$$\tan 0 < \tan(\alpha_i - \alpha_j) < \tan\left(\frac{\pi}{6}\right)$$

The last inequality implies that

$$0 < \frac{x_i - x_j}{1 + x_i x_j} < \frac{\sqrt{3}}{3}$$

which is what needed to be proven.

Problem 3

Prove the identity

$$\frac{a-b}{1+ab} + \frac{b-c}{1+bc} + \frac{c-a}{1+ca} = \frac{a-b}{1+ab} \cdot \frac{b-c}{1+bc} \cdot \frac{c-a}{1+ca}$$

Solution

Let us put $a = \tan\alpha$, $b = \tan\beta$, $c = \tan\gamma$, where $\alpha, \beta, \gamma \in \left(-\frac{\pi}{2}, \frac{\pi}{2}\right)$. Notice that

$$\frac{a-b}{1+ab} = \frac{\tan\alpha - \tan\beta}{1 + \tan\alpha\tan\beta} = \tan(\alpha - \beta)$$

$$\frac{b-c}{1+bc} = \frac{\tan\beta - \tan\gamma}{1 + \tan\beta\tan\gamma} = \tan(\beta - \gamma)$$

$$\frac{c-a}{1+ca} = \frac{\tan\gamma - \tan\alpha}{1 + \tan\gamma\tan\alpha} = \tan(\gamma - \alpha)$$

Therefore, the equality becomes

$$\tan(\alpha - \beta) + \tan(\beta - \gamma) + \tan(\gamma - \alpha) = \tan(\alpha - \beta) \cdot \tan(\beta - \gamma) \cdot \tan(\gamma - \alpha)$$

Let us put $x = \alpha - \beta$, $y = \beta - \gamma$, $z = \gamma - \alpha$. Notice that

$$x + y + z = (\alpha - \beta) + (\beta - \gamma) + (\gamma - \alpha) = 0$$

and it will be enough to prove that

$$\tan x + \tan y + \tan z = \tan x \cdot \tan y \cdot \tan z$$

Let us consider the following equality with $\tan z$:

$$\tan z = \tan\left(-x - y\right)$$

$$\tan z = -\tan\left(x + y\right)$$

$$\tan z = -\frac{\tan x + \tan y}{1 - \tan x \tan y}$$

$$\tan z \left(1 - \tan x \tan y\right) = -\left(\tan x + \tan y\right)$$

$$\tan z - \tan x \tan y \tan z = -\tan x - \tan y$$

$$\tan x + \tan y + \tan z = \tan x \tan y \tan z$$

which is what needed to be proven.

CHAPTER 49

Vectors

Vector is a mathematical concept used to take into account magnitude and direction. A vector is typically represented as an ordered pairs of numbers, called components, which provide information about how much the vector extends along each coordinate axis. Vectors possess numerous properties and are highly valuable tools in algebraic math olympiad problems. They offer a way to approach and solve problems that might be challenging or even impossible using alternative methods. Vectors often provide a geometric and algebraic framework that can lead to elegant and insightful solutions.

Vector

Vector is an ordered pair $\langle x, y \rangle$, where x and y are real numbers called the **components** of the vector.

The zero vector $\mathbf{0}$ is defined as $\mathbf{0} = \langle 0, 0 \rangle$.

Operations on Vectors

Given two vectors $\mathbf{a_1} = \langle x_1, y_1 \rangle$, $\mathbf{a_2} = \langle x_2, y_2 \rangle$, and a real number c. The operations of addition and subtraction of vectors and the multiplication of a vector by a number are defined as follows:

$$\mathbf{a_1} + \mathbf{a_2} = \langle x_1 + x_2, y_1 + y_2 \rangle$$
$$\mathbf{a_1} - \mathbf{a_2} = \langle x_1 - x_2, y_1 - y_2 \rangle$$
$$c\mathbf{a_1} = \langle cx_1, cy_1 \rangle$$

Magnitude of a Vector

The magnitude of the vector $\mathbf{a} = \langle x, y \rangle$ is defined as

$$|\mathbf{a}| = \sqrt{x^2 + y^2}$$

and represents the length of the corresponding directed segment.

Unit Vectors

Vector $\mathbf{e}$ is called a **unit vector** if its magnitude is equal to 1:

$$|\mathbf{e}| = 1$$

Linear Combination

If $\mathbf{e_1}$ and $\mathbf{e_2}$ are some noncollinear unit vectors, then any other vector $\mathbf{a}$ can be expressed as a **linear combination**:

$$\mathbf{a} = k_1\mathbf{e_1} + k_2\mathbf{e_2}$$

where k_1 and k_2 are some real numbers.

Dot Product

The dot product of the vectors $\mathbf{a_1} = \langle x_1, y_1 \rangle$ and $\mathbf{a_2} = \langle x_2, y_2 \rangle$ is the operation defined as

$$\mathbf{a_1} \cdot \mathbf{a_2} = x_1 x_2 + y_1 y_2$$

The dot product is related to the cosine of the angle ϕ between the vectors $\mathbf{a_1} = \langle x_1, y_1 \rangle$ and $\mathbf{a_2} = \langle x_2, y_2 \rangle$:

$$\mathbf{a_1} \cdot \mathbf{a_2} = |\mathbf{a_1}||\mathbf{a_2}| \cos \phi$$

Geometric Properties of Vectors

- The **directed segment** $\overrightarrow{AB}$ connecting the points $A(a_1, a_2)$ and $B(b_1, b_2)$ can be associated with the vector $\overrightarrow{AB} = \langle b_1 - a_1, b_2 - a_2 \rangle$.

- The **parallelogram law** states that if $ABCD$ is a parallelogram, then

$$\overrightarrow{AB} + \overrightarrow{AD} = \overrightarrow{AC}$$

- The **triangle law** states that if ABC is a triangle, then

$$\overrightarrow{AB} + \overrightarrow{BC} = \overrightarrow{AC}$$

Let us consider several problems.

Problem 1

Given 8 real numbers a, b, c, d, e, f, g and h. Prove that among the numbers $ac + bd$, $ae + bf$, $ag + bh$, $ce + df$, $cg + dh$, $eg + fh$, at least one number is nonnegative.

Solution

Let us assume that all mentioned numbers are negative.

Let us consider the vectors

$$\overrightarrow{OX_1} = \langle a, b \rangle$$
$$\overrightarrow{OX_2} = \langle c, d \rangle$$
$$\overrightarrow{OX_3} = \langle e, f \rangle$$
$$\overrightarrow{OX_4} = \langle g, h \rangle$$

Notice that the numbers $ac+bd$, $ae+bf$, $ag+bh$, $ce+df$, $cg+dh$, $eg+fh$ represent the following dot-products:

$$\overrightarrow{OX_1} \cdot \overrightarrow{OX_2} = ac + bd$$
$$\overrightarrow{OX_1} \cdot \overrightarrow{OX_3} = ae + bf$$
$$\overrightarrow{OX_1} \cdot \overrightarrow{OX_4} = ag + bh$$
$$\overrightarrow{OX_2} \cdot \overrightarrow{OX_3} = ce + df$$
$$\overrightarrow{OX_2} \cdot \overrightarrow{OX_4} = cg + dh$$
$$\overrightarrow{OX_3} \cdot \overrightarrow{OX_4} = eg + fh$$

Since we assumed that these values are negative it implies that all the angles $\angle X_i O X_j$ between the segments OX_i and OX_j are obtuse, which leads to a contradiction.

Problem 2

Let a, b, c, x, y, z be real numbers, such that

$$a^2 + x^2 = b^2 + y^2 = c^2 + z^2 = 1$$
$$(a + b)^2 + (x + y)^2 = (b + c)^2 + (y + z)^2 = (c + a)^2 + (z + x)^2 = 1$$

Show that

$$a + b + c = x + y + z$$

Solution

Let us consider the vectors

$$\overrightarrow{OA} = \langle a, x \rangle$$
$$\overrightarrow{OB} = \langle b, y \rangle$$
$$\overrightarrow{OC} = \langle c, z \rangle$$

From the first equation we can see that the vectors $\overrightarrow{OA}$, $\overrightarrow{OB}$, $\overrightarrow{OC}$ are unit vectors.

Let us now define the vectors

$$\overrightarrow{OD} = \langle a + b, x + y \rangle$$
$$\overrightarrow{OE} = \langle b + c, y + z \rangle$$
$$\overrightarrow{OF} = \langle c + a, z + x \rangle$$

From the second equation we can see that the vectors $\overrightarrow{OD}$, $\overrightarrow{OE}$, $\overrightarrow{OF}$ are also unit vectors and, therefore, the points A, B, C, D, E and F lie on the unit circle.

Notice that we also have

$$\overrightarrow{OA} + \overrightarrow{OB} = \langle a, x \rangle + \langle b, y \rangle = \langle a + b, x + y \rangle = \overrightarrow{OD}$$
$$\overrightarrow{OB} + \overrightarrow{OC} = \langle b, y \rangle + \langle c, z \rangle = \langle b + c, y + z \rangle = \overrightarrow{OE}$$
$$\overrightarrow{OC} + \overrightarrow{OA} = \langle c, z \rangle + \langle a, x \rangle = \langle c + a, z + x \rangle = \overrightarrow{OF}$$

This in particular implies that $OADB$ is a parallelogram. Furthermore, since $OA = OB$, then $OADB$ is a rhombus. Also, since $OA = OD$, then the triangles OAD and OBD are equilateral and $\angle AOD = \angle BOD = 60°$. Similarly, we can show that $\angle BOE = \angle COE = 60°$ and $\angle COF = \angle AOF = 60°$. This implies that the hexagon $ABCDEF$ is regular and $\angle AOB = \angle BOC = \angle COA = 120°$. Then $\overrightarrow{OD} = -\overrightarrow{OC}$ and $\overrightarrow{OA} + \overrightarrow{OB} = \overrightarrow{OD}$ imply that

$$\overrightarrow{OA} + \overrightarrow{OB} + \overrightarrow{OC} = \mathbf{0}$$
$$\langle a, x \rangle + \langle b, y \rangle + \langle c, z \rangle = \langle 0, 0 \rangle$$
$$\langle a + b + c, x + y + z \rangle = \langle 0, 0 \rangle$$

From here

$$a + b + c = x + y + z = 0$$

which is what needed to be proven.

Problem 3

The plane is tiled with congruent equilateral triangles of side 1. Let d_1 be the distance between some vertices A_1 and B_1 and d_2 be the distance between some vertices A_2 and B_2. Is it always possible to find the vertices A_3 and B_3, such that the distance d_3 between A_3 and B_3 satisfies $d_3 = d_1 \cdot d_2$.

Solution

Answer: yes, it is possible.

Let us consider two unit vectors e_1 and e_2 that form $120°$ and are collinear with the sides of the triangles. Notice that

$$e_1 \cdot e_1 = e_1^2 = 1$$

$$e_2 \cdot e_2 = e_2^2 = 1$$

The dot product of these vectors is

$$e_1 \cdot e_2 = |e_1||e_2|\cos 120° = -\frac{1}{2}$$

Let

$$\overrightarrow{A_1 B_1} = ae_1 + be_2$$

$$\overrightarrow{A_2 B_2} = ce_1 + de_2$$

for some integers a, b, c and d.

From here we have

$$d_1^2 = \overrightarrow{A_1 B_1} \cdot \overrightarrow{A_1 B_1}$$
$$= (ae_1 + be_2) \cdot (ae_1 + be_2)$$
$$= a^2 e_1^2 + 2abe_1 \cdot e_2 + b^2 e_2^2$$
$$= a^2 - ab + b^2$$

and

$$d_2^2 = \overrightarrow{A_2 B_2} \cdot \overrightarrow{A_2 B_2}$$
$$= (ce_1 + de_2) \cdot (ce_1 + de_2)$$
$$= c^2 e_1^2 + 2cde_1 \cdot e_2 + d^2 e_2^2$$
$$= c^2 - cd + d^2$$

Furthermore

$$d_1^2 \cdot d_2^2 = \left(a^2 - ab + b^2\right)\left(c^2 - cd + d^2\right)$$
$$= a^2 c^2 + b^2 d^2 - a^2 cd + a^2 d^2 - abc^2 + abcd - abd^2 + b^2 c^2 - b^2 cd$$
$$= (ac - bd)^2 - (ac - bd)(ad + bc - db) + (ad + bc - db)^2$$
$$= x^2 - xy + y^2$$

where $x = ac - bd$ and $y = ad + bc - db$.

From here we conclude that the vector

$$\overrightarrow{A_3 B_3} = xe_1 + ye_2$$

satisfies the conditions of the problem.

CHAPTER 50

Complex Numbers

Complex numbers extend the concept of real numbers by introducing a new element called the imaginary unit, denoted as i. It creates a number system that includes solutions to equations that cannot be solved using only real numbers. When the imaginary unit is combined with real numbers, it forms the complex number plane, which is a two-dimensional plane where the horizontal axis represents the real part of a complex number, and the vertical axis represents the imaginary part. Complex numbers are an important topic in algebraic math olympiad problems. They offer a way to approach and solve problems that might be challenging or even impossible using alternative methods and can lead to elegant solutions.

Complex number

Complex number z is defined as the number of the form $z = a + bi$, where a and b are real numbers and $i^2 = -1$. Geometrically, the complex number $a + bi$ can be associated with a point with coordinates (a, b) on the coordinate plane.

Real and Imaginary Parts

Given the complex number $z = a + bi$. Number a is called the **real part** of z and is denoted as $a = \Re(z)$. Number b is called the **imaginary part** of z and is denoted as $b = \Im(z)$.

Complex Conjugate

The complex number $\bar{z} = a - bi$ is called the **complex conjugate** for the complex number $z = a + bi$.

Operations on Complex Numbers

The operations of **addition**, **subtraction**, **multiplication** and **division** of the complex numbers $z_1 = a_1 + b_1 i$ and $z_2 = a_2 + b_2 i$ are defined as follows:

$$z_1 + z_2 = (a_1 + a_2) + (b_1 + b_2)\, i$$

$$z_1 - z_2 = (a_1 - a_2) + (b_1 - b_2)\, i$$

$$z_1 \cdot z_2 = (a_1 a_2 - b_1 b_2) + (a_2 b_1 + a_1 b_2)\, i$$

$$\frac{z_1}{z_2} = \left(\frac{a_1 a_2 + b_1 b_2}{a_2^2 + b_2^2} \right) + \left(\frac{a_2 b_1 - a_1 b_2}{a_2^2 + b_2^2} \right) i$$

Argument of a Complex Number

The **argument** ϕ of the complex number $z = a + bi$ is defined as follows:

- If $a = 0$ and $b = 0$, then ϕ is undefined.
- If $a = 0$ and $b > 0$, then $\phi = \frac{\pi}{2}$.
- If $a = 0$ and $b < 0$, then $\phi = \frac{3\pi}{2}$.
- If $a > 0$, then $\phi = \tan^{-1}\left(\frac{b}{a}\right)$.
- If $a < 0$, then $\phi = \pi + \tan^{-1}\left(\frac{b}{a}\right)$.

Magnitude of a Complex Number

The **magnitude** $|z|$ of the complex number $z = a + bi$ is defined as

$$|z| = \sqrt{z \cdot \bar{z}} = \sqrt{a^2 + b^2}$$

Trigonometric Form of a Complex Number

The **trigonometric form** of the complex number z is

$$z = |z| \left(\cos \phi + i \sin \phi \right)$$

where $|z|$ and ϕ are the magnitude and the argument of z, respectively.

Multiplication and Division in Trigonometric Form

The multiplication and division of the complex numbers $z_1 = |z_1|(\cos \phi_1 + i \sin \phi_1)$ and $z_2 = |z_2|(\cos \phi_2 + i \sin \phi_2)$ can be done as follows:

$$z_1 \cdot z_2 = |z_1||z_2| \left(\cos \left(\phi_1 + \phi_2 \right) + i \sin \left(\phi_1 + \phi_2 \right) \right)$$

$$\frac{z_1}{z_2} = \frac{|z_1|}{|z_2|} \left(\cos \left(\phi_1 - \phi_2 \right) + i \sin \left(\phi_1 - \phi_2 \right) \right)$$

De Moivre's Formula

The exponentiation of the complex number $z = |z| \left(\cos \phi + i \sin \phi \right)$ can be done as:

$$z^n = \cos \left(\phi n \right) + i \sin \left(\phi n \right)$$

Roots of a Complex Number

The roots z_k of order n of the complex number $z = |z| \left(\cos \phi + i \sin \phi \right)$ are given by the formula:

$$z_k = \sqrt[n]{|z|} \left(\cos \frac{\phi + 2\pi k}{n} + i \sin \frac{\phi + 2\pi k}{n} \right), \text{ where } k = 0, 1, \ldots, n - 1.$$

Let now us consider several problems.

Problem 1

Solve the system of equations

$$\begin{cases} x + \dfrac{3x - y}{x^2 + y^2} = 3 \\[4mm] y - \dfrac{x + 3y}{x^2 + y^2} = 0 \end{cases}$$

Solution

Let us start by multiplying the second equation by i:

$$yi - \frac{xi + 3yi}{x^2 + y^2} = 0$$

Let us now add it with the first equation:

$$x + \frac{3x - y}{x^2 + y^2} + yi - \frac{xi + 3yi}{x^2 + y^2} = 3$$

$$(x + yi) + \frac{3x - y - xi - 3yi}{x^2 + y^2} = 3$$

$$(x + yi) + \frac{3(x - yi) + (-y - xi)}{x^2 + y^2} = 3$$

$$(x + yi) + \frac{3(x - yi) - (x - yi)i}{x^2 + y^2} = 3$$

Let us now make the substitution $z = x + yi$. Then $\bar{z} = x - yi$ and

$$z \cdot \bar{z} = x^2 + y^2$$

Therefore, the equation becomes

$$z + \frac{3\bar{z} - \bar{z}i}{z\bar{z}} = 3$$

$$z + \frac{\bar{z}(3 - i)}{z\bar{z}} = 3$$

$$z + \frac{3 - i}{z} = 3$$

$$z^2 + 3 - i = 3z$$

$$z^2 - 3z + 3 - i = 0$$

We can solve the last equation using the Quadratic Formula:

$$z = \frac{3 \pm \sqrt{4i - 3}}{2} = \frac{3 \pm \sqrt{(2i+1)^2}}{2} = \frac{3 \pm (2i+1)}{2}$$

which implies that $z = 2 + i$ or $z = 1 - i$. This in turn gives the following pairs (x, y) of the solutions of the initial equation: $(2, 1)$, $(1, -1)$.

Problem 2

Prove the identity

$$\cos\left(\frac{2\pi}{7}\right) + \cos\left(\frac{4\pi}{7}\right) + \cos\left(\frac{6\pi}{7}\right) = -\frac{1}{2}$$

Solution

Let us consider the complex number

$$z = \cos\left(\frac{2\pi}{7}\right) + i\sin\left(\frac{2\pi}{7}\right)$$

Notice that by the De Moivre's Formula

$$z^2 = \cos\left(\frac{4\pi}{7}\right) + i\sin\left(\frac{4\pi}{7}\right)$$

$$z^3 = \cos\left(\frac{6\pi}{7}\right) + i\sin\left(\frac{6\pi}{7}\right)$$

and it will be enough to prove that

$$\Re\left(z + z^2 + z^3\right) = -\frac{1}{2}$$

Let us use the De Moivre's Formula to find the value of z^7:

$$z^7 = \cos(2\pi) + i\sin(2\pi) = 1$$

Then

$$0 = z^7 - 1 = (z - 1)\left(z^6 + z^5 + z^4 + z^3 + z^2 + z + 1\right)$$

This implies that

$$z^6 + z^5 + z^4 + z^3 + z^2 + z + 1 = 0$$

Let us now consider the equality $z^7 = 1$. It implies that

$$z^6 = \frac{1}{z} = \overline{z}$$

$$z^5 = \frac{1}{z^2} = \overline{z^2}$$

$$z^4 = \frac{1}{z^3} = \overline{z^3}$$

Therefore

$$z^6 + z^5 + z^4 + z^3 + z^2 + z + 1 = 0$$

$$\overline{z} + \overline{z^2} + \overline{z^3} + z^3 + z^2 + z + 1 = 0$$

$$\overline{\left(z + z^2 + z^3\right)} + \left(z + z^2 + z^3\right) + 1 = 0$$

$$2\Re\left(z^3 + z^2 + z\right) + 1 = 0$$

$$2\Re\left(z^3 + z^2 + z\right) = -1$$

$$\Re\left(z + z^2 + z^3\right) = -\frac{1}{2}$$

which is what needed to be proven.

Problem 3

Let a, b, c, x, y, z be real numbers, such that

$$a^2 + x^2 = b^2 + y^2 = c^2 + z^2 = 1$$

$$(a + b)^2 + (x + y)^2 = (b + c)^2 + (y + z)^2 = (c + a)^2 + (z + x)^2 = 1$$

Show that

$$a^2 + b^2 + c^2 = x^2 + y^2 + z^2$$

Solution

Let us introduce the following complex numbers:

$$z_1 = a + xi$$

$$z_2 = b + yi$$

$$z_3 = c + zi$$

Notice that the conditions of the problem imply that

$$|z_1| = |z_2| = |z_3| = 1$$

and

$$|z_1 + z_2| = |z_2 + z_3| = |z_3 + z_1| = 1$$

Let us prove that

$$z_1 + z_2 + z_3 = 0$$

Let the point O be the origin, the points A_1, A_2, A_3 represent the complex numbers z_1, z_2, z_3, and the points B_1, B_2, B_3 represent the complex numbers $z_2 + z_3$, $z_3 + z_1$, $z_1 + z_2$, respectively. Then the triangles OA_iB_j are equilateral and all segments A_iB_j are congruent for all $i \neq j$. From here $A_1B_3A_2B_1A_3B_2$ is a regular hexagon inscribed into a unit circle. Notice that all of its vertices can be obtained from one vertex by 60° rotation, i.e. by the multiplication by the complex number

$$\beta = \cos\left(\frac{\pi}{3}\right) + i\sin\left(\frac{\pi}{3}\right)$$

Let us consider the sum of all the complex numbers that represent the vertices of the regular hexagon $A_1B_3A_2B_1A_3B_2$. Taking into account that $\beta^6 = 1$ we have

$$3(z_1 + z_2 + z_3) = z_1 + z_1\beta + z_1\beta^2 + z_1\beta^3 + z_1\beta^4 + z_1\beta^5$$

$$3(z_1 + z_2 + z_3) = z_1\left(1 + \beta + \beta^2 + \beta^3 + \beta^4 + \beta^5\right)$$

$$3(z_1 + z_2 + z_3) = z_1\left(\frac{1 - \beta^6}{1 - \beta}\right)$$

$$3(z_1 + z_2 + z_3) = 0$$

$$z_1 + z_2 + z_3 = 0$$

Let us now prove that

$$z_1^2 + z_2^2 + z_3^2 = 0$$

Indeed, we have

$$z_1^2 + z_2^2 + z_3^2 = (z_1 + z_2 + z_3)^2 - 2(z_1z_2 + z_2z_3 + z_3z_1)$$

$$= -2(z_1z_2 + z_2z_3 + z_3z_1)$$

$$= -2z_1z_2z_3\left(\frac{1}{z_3} + \frac{1}{z_1} + \frac{1}{z_2}\right)$$

$$= -2z_1z_2z_3\left(\overline{z_3} + \overline{z_1} + \overline{z_2}\right)$$

$$= -2z_1z_2z_3\overline{(z_1 + z_2 + z_3)}$$

$$= 0$$

On another hand we have

$$z_1^2 + z_2^2 + z_3^2 = a^2 - x^2 + b^2 - y^2 + c^2 - z^2$$

Since $z_1^2 + z_2^2 + z_3^2 = 0$, then

$$a^2 + b^2 + c^2 = x^2 + y^2 + z^2$$

as desired.[1]

[1] We solved a similar problem in a different way in Chapter 49 "Vectors".

Made in the USA
Monee, IL
08 January 2025

76352781R00148